Geoplanet: Earth and Planetary Sciences

For further volumes:
http://www.springer.com/series/8821

Dorota Mirosław-Świątek
Tomasz Okruszko
Editors

Modelling of Hydrological Processes in the Narew Catchment

 Springer

Dorota Mirosław-Świątek
Department of Hydraulic Engineering
Warsaw University of Life Sciences
SGGW
Nowoursynowska 166
02-787 Warsaw
Poland
e-mail: dorotams@levis.sggw.pl

Tomasz Okruszko
Department of Hydraulic Engineering
Warsaw University of Life Sciences
SGGW
Nowoursynowska 166
02-787 Warsaw
Poland
e-mail: T.Okruszko@levis.sggw.pl

The GeoPlanet: Earth and Planetary Sciences Book Series is in part a continuation of Monographic Volumes of Publications of the Institute of Geophysics, Polish Academy of Sciences, the journal published since 1962 (http://pub.igf.edu.pl/index.php).

ISSN 2190-5193
ISBN 978-3-642-26958-5
DOI 10.1007/978-3-642-19059-9
Springer Heidelberg Dordrecht London New York

e-ISSN 2190-5207
ISBN 978-3-642-19059-9 (eBook)

Series Editors

Managing Editors

Advisory Board

Preface

Modeling of hydrological conditions in lowland rivers has a long tradition and includes a number of methods which aim at mimic of hydrological processes. Most of the models are developed for the certain purposes like river basin planning, operation of hydraulic structures, environmental impact assessments of particular investments or identification of hydrological properties of different habitats or ecosystems. The purpose of the model shapes its structure, resolution and dynamics.

The aim of this book is a comparison of a number of specific models and modeling methods used in the Narew River basin. There is a strong advantage of comparison of models built for the same area as it illustrates better the advantages and constraints of particular approaches in models used for different purposes. Moreover, the Narew River basin is on one hand a typical river system of Central European Lowlands, but on the other hand it is a basin very rich in valuable habitats and water dependent ecosystems. This means that modeling efforts are also important for recognition and preservation of those values for the future generations.

The Narew River Basin is situated in the north-eastern part of Poland. The Narew River is the fifth largest in the country with regard to river length (484 km) and the size of the basin (ca. 28,000 km^2 before joining the Bug river). The entire basin, except for the most upper part (ca. 1,200 km^2), is located in Poland, but the head catchment is located in Belarus. The basin is developed in the reach of two last glaciations periods: Riss and Wuerm. From north to south the basin includes the following sequence of glacial landscapes types: moraine lake district, outwash plains, ice-marginal river valleys and moraine hills. Sandy soils of various types predominate. During the Holocene the main valleys have been filled up with mesotrophical-eutrophic peat layers which still are partly undrained at present.

Poland is located in a temperate climatic zone in which marine and continental air masses collide. This climate type features characterize also the Narew River Basin. However, due to its location in northeastern Poland the impact of continental air is more visible in the basin compared to the rest of the country. The yearly average temperature is 7.2°C and precipitation equals 617 mm. The river network of the basin is fully developed and rich in tributaries, which mostly originate in the postglacial lakes located in the northern part of the basin. In the lake district region there are more than 500 lakes greater then 1 ha. A few irrigation and navigation channels creates interconnections between lakes and the river network. The flow regime is typical for the lowland rivers in this part of Europe with peak flows after snow melting and regularly appearing low flow periods in the fall of the summer. The yearly average flow recorded in the most downstream gauging station—Zambski Koscielne (before confluence of the Bug and Narew rivers in the artificial lake) equals to 147 m^3 s^{-1}, when the average yearly minimum is 55 m^3 s^{-1}.

The area of Narew River Basin belongs to poorly populated in the scale of the country. The estimated number of inhabitants of the region is about 1.5 million. On average 55 inhabitants occur for square kilometre of the basin area, while in the rest of the country average population density is more than twice as large, being up to 123 inhabitants per square kilometre. More than half of the population (60%) lives in the cities and towns. The biggest city on the analyzed area is Białystok, the capital city of Podlaskie voivodeship, with 285,000 inhabitants. The other cities are decidedly smaller and none of them exceeds 70,000 inhabitants. All cities have sanitary sewerage systems, transporting effluents to wastewater treatment plants, and storm drainage systems that drain off precipitation water to the nearest receiving water. In most of the cities sewerage network is distributive.

Narew River basin is an agricultural region, with a small degree of industrialization and no heavy industry. Existing production is connected with agriculture and forestry, and is based on local raw materials, which are mainly: milk, meat, cereals, vegetables, fruits and wood. Industries that are developing are mainly agricultural, food and timber processing, and recently tourism. Wastewater from enterprises located in the cities in most cases discharge through the municipal main sewerage system and thence to wastewater treatment plants. Enterprises that are disperse through the basin area have their own effluent treatment. Agricultural land dominates the basin, covering almost 55% of its area. The upland of basin area is mainly used for arable land, the valleys are used as pastures and grasslands. The forestation ratio of the Narew River Basin is slightly over 32%, which somewhat exceeds the entire country average. The largest and compact forest complexes are located in the north, west and east parts of the basin.

The basin is rich in nature areas and resources. The Narew and the Biebrza river valleys are among Europe's last active, regularly flooded riverine valleys. Until now, a considerable part of this area had been utilized for the purposes of extensive (environmentally sound) agricultural practices, thanks to which it still boasts wet meadows of a significant biodiversity value. Additionally in the south-eastern part of the basin, there are number of alder carrs which are groundwater fed. All those habitats are protected in the form of national parks. The vegetation cover of the Narew river valley comprises of several dominating communities such as: reed beds, sedge communities, fens, humid meadows, single mown meadows and herbaceous communities. Twelve water dependent plant communities have been recorded in the valley and listed in the Council Directive 92/43/EEC of

21st May 1992 (Habitat Directive). These are the following: Alkaline fens (54.2), Lowland hay meadows (38.2), Molinia meadows (37.31), Eutrophic tall herbs (37.7), Quaking mires (54.5), Residual alluvial forests (44.3) and Oak-hornbeam forests (41.24).

The Narew and the Biebrza river valleys are also very important areas from wild life existence and protection point of view, especially with regard to bird communities. The valley meets all criteria of Bird Life International. The area is a nesting ground for 90 endangered species out of which 52 are wetland birds. In particular, it is a breeding ground for three species threatened with worldwide extinction and a breeding area for 1% or more of the European population of at least ten species of wetland birds. On the local scale, it is as well very special area being one of the ten most important refuges in Poland for at least 22 bird species. The basin's natural value and importance is emphasised by the existence of three large national parks (ca. 750 km^2) and a vast number of other protected areas (Natura 2,000 sites ca 2,500 km^2, landscape parks ca. 1,300 km^2, strict reserves ca. 170 km^2).

The modeling of hydrological process in the Narew River Basin has been described in eight papers which form a content of this book. Six of them aim at description of the surface waters, the remaining two focus on the groundwater. The first paper of Kadłubowski et al. describes the rainfall/snowmelt-runoff model for the Upper Narew River basin. This model is a part of the Integrated Hydrological Monitoring and Forecasting System for the Vistula River Basin. The idea of the model and the main procedures are described. The second paper of Piniewski and Okruszko describes the application of the hydrological component of the catchment model, Soil and Water Assessment Tool (SWAT) in the whole Narew basin. The main objective was to perform a multi-site calibration and validation of SWAT using daily observed flows from 23 gauging stations as well as to assess the model's capability to perform reliable simulations at spatial scales that were smaller than in the calibration phase. Porretta-Brandyk et al. in the third paper deal with the Upper Basin of the Biebrza River. For the catchment areas of two small tributaries WetSpa model was applied for runoff simulation based on soil-atmosphere-plant mass balance at a catchment scale. The simulated hydrological processes included: precipitation, evapotranspiration, plant canopy interception, soil interception, infiltration and capillary rise, ground water flow. The last paper in this group (Banaszuk et al) analyses the subcatchment inflow using End-Member Mixing Analysis (EMMA). The authors identify flow paths and source areas controlling river chemistry during a snow melt induced spring high flow event. EMMA using Ca^{2+}, Mg^{2+}, Cl^-, and H^+ showed that stream chemistry could be explained as a three-component mixture of overland flow, shallow groundwater and soil solution from arable soils along the stream margin and deeper groundwater. The temporal variability in the flow pathways and solute sources during snowmelt were explained.

The next two papers deal with the groundwater models: one in the watershed area; second in the groundwater fed mired named Red Bog. The main goal of the first paper (Mioduszewski et al.) is to explore by hydrological modeling the groundwater system on both sides of the watershed boundary between the rivers Narew and Supraśl and the influence of two nearby peat mines. An integrated model of the area was set up using the SIMGRO program. The aim of the model simulations was to estimate the historic hydrological situation of the area. In the second paper of Grygoruk et al, a three dimensional finite-difference steady-state groundwater model was applied to analyze the groundwater flow system of the Middle Biebrza Basin. The study contains analysis of hydrogeological and morphological outline of the area, as well as the description of developed groundwater model including conceptual model description, model calibration and sensitivity analysis of parameters.

The last two papers focus on surface water management issues using the hydrodynamic models for the Upper Narew valley. Kiczko and Napiórkowski present a Multiple Criteria Decision Support System (DSS), based on the Aspiration Reservation method, for the optimal management of the Siemianówka reservoir. The proposed DSS makes it possible to find a trade-off between different reservoir goals, such as: agriculture, fisheries, energy production and wetland ecosystems. The control problem is solved using Receding Horizon Optimal Control Technique. Kubrak et al. analyze the results of the dam break wave forecast in the valley of the Narew River, after the failure of the Siemianówka Reservoir earth dam, are presented. The velocity of the wave-front propagation in the Narew Valley and discharge characteristics as well as the inundation areas are calculated using 1-D hydrodynamic model.

In our view, the papers in the present volume of Geoplanet once again support the notion that only the cooperation of experts from different modeling groups can increase the quality of models and find the gaps in our understanding of hydrological processes.

Dorota Mirosław-Świątek
Tomasz Okruszko

Contents

Operational Rainfall/Snowmelt-Runoff Model for Upper Narew River

Andrzej Kadłubowski, Małgorzata Mierkiewicz
and Halina Budzyńska

Abstract This paper presents results of the rainfall/snowmelt-runoff model for the Upper Narew River basin. This model is a part of the Integrated Hydrological Monitoring and Forecasting System for the Vistula River Basin. The idea of the model and main procedures are described. Several remarks and experiences of the calibration process are presented.

1 Introduction

In 1996 the Institute of Meteorology and Water Management in Warsaw (IMGW) and the Swedish Meteorological and Hydrological Institute in Norrköping, Sweden (SMHI) started to co-operate in the project Integrated Hydrological Monitoring and Forecasting System (IHMS) for the Vistula River Basin (Mierkiewicz et al. 1999). The System was developed based on the Swedish rainfall/snowmelt-runoff model, HBV and the flow routing hydrodynamic model, HD, in use at the Polish Institute (Lindel et al. 1997). During few months the System was developed for the lower and middle Vistula river basin and finally it was applied in the operational hydrological service at the IMGW.

Main tributaries or their upper parts were modeled by the HBV model. This model was also set-up and calibrated for the Upper Narew River.

A. Kadłubowski (✉) · M. Mierkiewicz · H. Budzyńska
Institute of Meteorology and Water Management, Cracow Center of Flood Modeling,
61 Podleśna Str., 01-673 Warsaw, Poland
e-mail: Andrzej.Kadlubowski@imgw.pl

D. Mirosław-Świątek and T. Okruszko (eds.), *Modelling of Hydrological Processes in the Narew Catchment*, Geoplanet: Earth and Planetary Sciences,
DOI: 10.1007/978-3-642-19059-9_1, © Springer-Verlag Berlin Heidelberg 2011

2 Rainfall/Snowmelt-Runoff Model HBV

Originally the rainfall-runoff model HBV was developed for runoff simulations and hydrological forecasting in the early seventies [for inflow forecasting to hydropower reservoirs in Scandinavian catchments (Berström 1976)]. It is characterized as a semidistributed conceptual model with moderate demands on input data. The model is usually run with daily values of rainfall and air temperature and monthly estimates of potential evapotranspiration. The model components clearly represent individual hydrological processes. It contains routines for snow accumulation and melting, soil moisture calculation, runoff generation and a simple routing procedure (Fig. 1). The catchment can be divided into sub-basins. Each subbasin is then divided into zones according to altitude, lake area and vegetation.

2.1 Main Procedures of the HBV Model

2.1.1 Snowmelt and Accumulation

Snowmelt is calculated separately for each elevation and vegetation zone according to the degree-day equation:

$$Q_m(t) = \text{CFMAX} \cdot (T(t) - TT) \tag{1}$$

where Q_m, snowmelt; CFMAX, degree-day factor; T, zone temperature; TT, temperature limit for snow/rain.

The threshold temperature is usually used to decide whether precipitation is rainfall or snowfall. It is possible to have different threshold for accumulation and melting. It is also possible to use a threshold interval for a rain and snow mix.

Because of the porosity of the snow some rain and meltwater can be retained in the pores. A retention capacity of 10% of the snowpack water equivalent is assumed in the model. Only after the retention capacity has filled, meltwater will be released from the snow. The snow routine also has a general snowfall correction factor which adjusts for systematic errors in calculated snowfall and winter evaporation.

2.1.2 Soil Moisture

The soil moisture routine is the main part controlling runoff formation. Soil moisture dynamics are calculated separately for each elevation and vegetation zone. The rate of discharge of excess water from the soil is related to the precipitation and the relationship depends upon the computed soil moisture storage, the maximum soil moisture content (FC) and the empirical parameter BETA, as

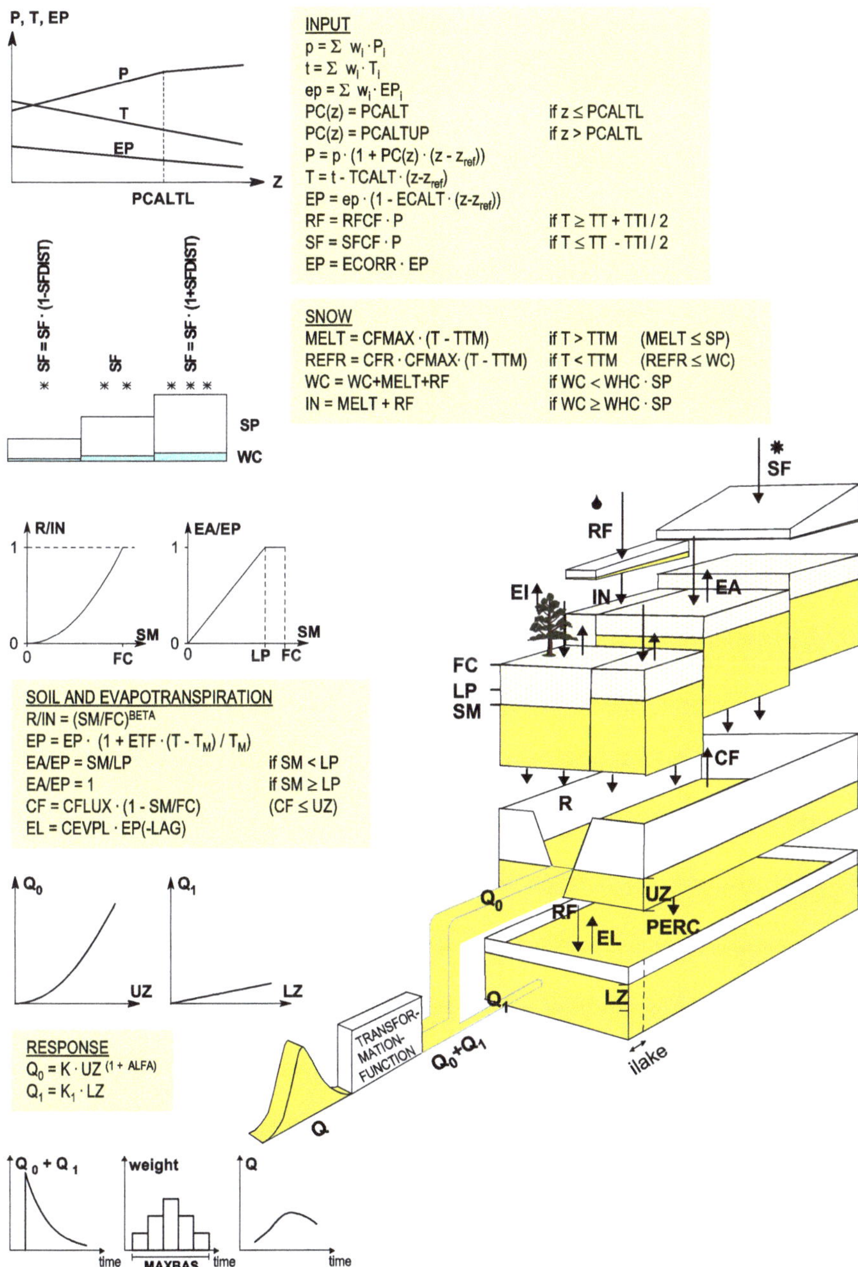

Fig. 1 Schematic structure of one subbasin in the HBV model, with routines for snow (*top*), soil (*middle*) and response (*bottom*) (*bold letters* in legend indicate that the parameter is normally calibrated). Detailed description of variables, parameters and constants can be found in (Lindel et al. 1997)

given in Eq. 2. Rain or snowmelt generate small contributions of excess water from the soil when the soil is dry and large contributions when conditions are wet.

$$Q_s(t) = \left(\frac{SM(t)}{FC}\right)^{BETA} \cdot P(t) \tag{2}$$

where Q_s, excess water from soil; SM, soil moisture storage; FC, maximum soil moisture content; P, zone precipitation; BETA, empirical coefficient.

The specified input potential evapotranspiration value can be corrected by a general evaporation correction factor and by altitude with an elevation correction factor. It can also be modified according to actual temperature related to normal temperature values.

The actual evapotranspiration is computed as a function of the potential evapotranspiration and the available soil moisture (Eq. 3):

$$EA(t) = \begin{cases} \dfrac{EP \cdot SM(t)}{LP} & \text{if} \quad SM \leq LP \\[2ex] EP & \text{if} \quad SM > LP \end{cases} \tag{3}$$

where EA, actual evapotranspiration; EP, zone potential evapotranspiration; LP, limit for potential evapotranspiration.

2.1.3 Runoff Response

Excess water from the soil and direct precipitation over open water bodies in the catchment area generate runoff from the response tanks according to Eqs. 4 and 5.

$$Q_0(t) = K \cdot UZ^{(1+ALFA)} \tag{4}$$

$$Q_1(t) = K_1 \cdot LZ \tag{5}$$

where Q_0, runoff generation from upper response tank; Q_1, runoff generation from lower response tank; K_0, K_1, recession coefficients; UZ, storage in upper response tank; LZ, storage in lower response tank; ALFA, response box parameter.

In order to account for the damping of the generated flood pulse $(Q = Q_0 + Q_1)$ in the river, a simple routing transformation is made. This is a filter with a triangular distribution of weights. There is also an option of using the Muskingum routing routine to account for the river flow hydraulics.

Lakes in the subbasins are included in the lower response tank, but they can also be modeled explicitly by a storage discharge relationship. This is accomplished by subdivision into subbasins defined by the outlet of major lakes. The use of an explicit lake routing routine has also proved to simplify the calibration of the recession parameters of the model, as most of the damping is accounted for by the lakes.

2.2 Model Calibration

Usually 5–10 years of records are sufficient for a stable model calibration for all kinds of applications. It is important that the records include a variety of hydrological events, so that the effect of all subroutines of the model can be discerned.

The HBV model, in its simplest form with only one subbasin and one type of vegetation, has altogether 12 free parameters (see the legend to Fig. 1, parameters are marked with bold letters). Parameter values are integrated and specific for each catchment and can not easily be obtained from point measurements in the field. Calibration of the model is usually made by a manual trial and error technique, during which relevant parameter values are changed until an acceptable agreement with observations is obtained. The judgment of the performance is also supported by statistical criteria, normally the R^2 value of model fit, $\approx$ explained variance (Nash and Sutcliffe 1970):

$$R^2 = \frac{\sum \left(\overline{Q}_o - Q_o\right)^2 - \sum \left(Q_c - Q_o\right)^2}{\sum \left(\overline{Q}_o - Q_o\right)^2} \qquad (6)$$

where Q_o, observed runoff; $\overline{Q}_o$, mean of observed runoff; Q_c, computed runoff.

R^2 has a value of 1.0, if the simulated and the recorded hydrographs agree completely, and 0 if the model only manages to produce the mean value of the runoff record. Another useful tool for the judgment of model performance is a graph of the accumulated difference between the simulated and the recorded runoff. This graph reveals any bias in the water balance and is often used in the initial stages of calibration.

2.3 The Forecasting Procedure

The HBV model can be used for short range forecasting (3-days ahead in IMGW). For the period previous to forecasting time the model is run on observed data and simulation results are compared with observations. If there is a discrepancy between the computed and observed hydrographs during the last days of run, updating of the model should be considered. The HBV model is usually updated either manually or automatically by an iterative procedure during which the input data a few days prior to the day of forecast is adjusted. In the automatic updating routine introduction of updating limits help to avoid unrealistic result.

In forecasting period a meteorological forecast is used as input, and there is a possibility to use alternative precipitation and temperature sequences in the same run. This is often desirable due to the often low accuracy of quantitative meteorological forecasts, especially as concerns precipitation.

Table 1 Climate and hydrological characteristics of the Upper Narew River basin (upstream of the hydrological station Wizna)

River length (km)	238.1
Area (km^2)	14307.7
Mean annual rainfall (mm)	600–650
Mean annual air temperature (°C)	5.5–7.0
Numbers of days with snow cover	80–120
Probability of snow winter (%)	40–60
Numbers of days with frost	About 100
Numbers of days with ground frost	Over 130
Mean summer and winter outflow (% of annual outflow)	38.5 (s)
	61.5 (w)
High flow (m^3/s)	992
Low flow (m^3/s)	10.0
Amplitude of water stage (m)	2.2–4.5
Period of ice phenomena	XI–III
Numbers of days with ice phenomena	Over 90
Mean slope of the river bad (‰)	0.25–0.05

3 Upper Narew River Catchment: Climate and Hydrology

The Upper Narew River basin is situated in North-Eastern part of Poland. The river has a hydrological regime typical to most of the East-European Rivers with the maximum flood during spring, in March and April. During summer and autumn the water level and the discharge is rather low with periodical increases due to rainfall. The arctic air is coming for few days in the winter causing very frosty days with snow-storm. In spring and autumn ground frost is observed. Characteristics of climate and hydrology of Upper Narew River are presented in Table 1.

4 The HBV Model for the Upper Narew River

In the HBV model setup the district Upper Narew was divided into 7 subbasins (Table 2) based on the location of discharge stations (Fig. 2). Runoff data from these stations are used for calibration of the model parameters.

4.1 Input Data Analysis

Precipitation is the source of stream flow generation and consequently the most important input data to the HBV model. Furthermore temperature data and long term estimates of potential evapotranspiration (e.g., monthly sums) are needed as input.

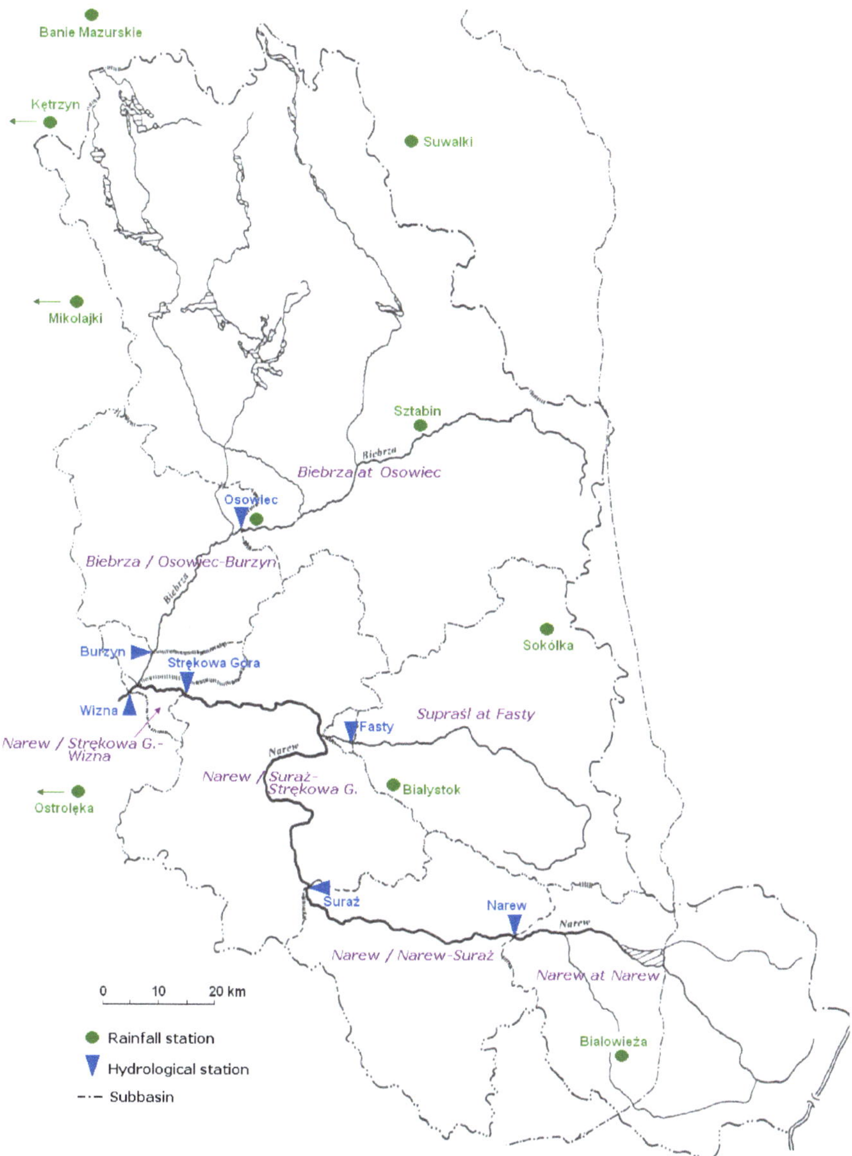

Fig. 2 Subbasin division of Upper Narew River district

A control of the homogeneity of the precipitation and discharge stations has been accomplished with the double mass technique. This technique takes advantage of the fact that the mean accumulated precipitation for the number of gauges is not very sensitive to changes at individual stations because many of the errors compensate each other. However the cumulative curve for a single gauge is

Table 2 Upper Narew River district and subbasins

District	Number of subbasins	Area (km^2)	Forest (%)	Open field (%)
Upper Narew	7	14307.7	35.7	64.3
Subbasins (River)				
Narew (Narew)	1	1978.0	62.2	37.8
Suraż (Narew)	2	1398.5	23.9	76.1
Fasty (Supraśl)	3	1816.6	54.1	45.9
Strękowa Góra (Narew)	4	1987.1	28.2	71.8
Osowiec (Biebrza)	5	4365.1	27.3	72.7
Burzyn (Biebrza)	6	2535.3	28.4	71.6
Wizna (Narew)	7	226.3	41.7	58.3

Table 3 Meteorological stations for Upper Narew River modeling

Meteorological station	Meteorological parameters	Mean annual precipitation (mm)
Kętrzyn	P, T	578
Mikołajki	P, T	602
Suwałki	P, T	598
Białowieża	P	660
Sztabin	P	597
Osowiec	P	549
Sokółka	P	663
Ostrołęka	P, T	581
Białystok	P, T	589
Banie Mazurskie	P	703

P precipitation, *T* air temperature

immediately affected by a change at the station. If the double mass curve has a change in slope at some point in time it indicates a break in homogeneity. A jag in the double mass curve can be caused by missing values at the observed station or by seasonal differences in the precipitation pattern. The slope of the curve is proportional to the intensity, i.e., if records from one station have similar mean as mean of the remaining stations, the curve follows the diagonal. If the station records more the slope will be steeper and if it records less, the double mass curve will lie below the diagonal. Finally chosen precipitation stations are presented in Table 3. Average precipitation for each subbasin was calculated based on Thiessen method.

In Table 4 the influence of anthropopression for discharge stations in Upper Narew River district is marked. Due to increase in population, growing economy, changes in land use and industrial development the runoff regime has been changed. Natural processes also affect discharges. In winter, when there are ice phenomenon, in summer, when the intensive vegetation in rivers is observed, the special coefficients for discharge have to be used but values of them are not constant and they are estimated subjectively.

Table 4 Water regime for discharge stations in Vistula river basin

Discharge station (river)	Area (km^2)	Mean discharge (m^3/s)	Mean unit outflow (l/s, km^2)
Narew (QN) (Narew)	1978.0	9.81	4.95
Suraż (N) (Narew)	3376.5	15.5	4.59
Fasty (C) (Supraśl)	1816.6	8.73	4.81
Strękowa Góra (C) (Narew)	7180.6	33.3	4.64
Osowiec (N) (Biebrza)	4365.1	22.4	5.13
Burzyn (N) (Biebrza)	6900.4	33.6	4.87
Wizna (N) (Narew)	14307.7	67.9	4.75

Type of water regime: N natural, QN quasi-natural, C changed

4.2 Calibration

Calibration was carried out against historical runoff data, with various periods (wet and dry). It was made by a manual trial and error technique, during which relevant parameter values are changed until an acceptable agreement with observations is obtained.

Calibration was carried out for different period, when the big spring flood (e.g., 1979) and very long hydrological drought (1982–1995) has happened. Verification has been made for another period, including years after 2000. In some causes recalibration of model parameters was needed.

Due to severe inhomogeneities in data calibration was not easy. In some cases in order to make better calibration the precipitation weights were changed with comparison to original values or some stations were eliminated.

In second part of 80-thies and in 90-thies hydrological drought was observed in Poland. During this period for many meteorological stations the precipitation deficit was bigger than mean annual sum of precipitation. This effect was reflected in observed hydrographs. In 90-thies for many gauge stations outflow was very low and water stages were lower than ever in history of observation. For few discharge stations systematic deviation was noticed in HBV model simulation. It was characteristic that since 1986 the accumulated difference has increased. In order to eliminate this deviation the precipitation correction factor lower than 1 has been used from this year.

Range of the most important model parameter values for selected subbasins in the Upper Narew River district is presented in Table 5.

4.3 Model Result

In calibration the R^2 value of model fit together with accumulated difference between the simulated and observed runoff was used to check model performance. This two hydrograph were also visually inspected, taking into account hydrographs shape and time accordance.

Table 5 Range of the most important model parameter values for the calibrated Upper Narew River subbasins

Parameter	Value	Function
Snow routine		
SFCF	1.0–1.7	Snowfall correction factor
TT	−1.0–0.5	Threshold temperature for snowmelt
CFmax	3.0–7.0	Degree-day factor
Soil routine		
FC	150.0–450.0	Field capacity
LP	0.35–0.90	Limit for potential evapotransporation
Beta	0.85–2.50	Empirical coefficient
Response routine, upper zone		
Cflux	0.0–1.2	Capillary flux
Alfa	0.2–1.5	Recession parameter
HQ	0.60–2.32	Peak flow level
KHQ	0.008–0.500	Recession at HQ
Response routine, lower zone		
Perc	0.08–5.00	Ground water percolation
K4	0.001–0.300	Base flow recession parameter

Table 6 R^2 values and accumulated differences (Accdiff) for the modeled outflow of the Upper Narew River subbasins (80-, 90-thies)

District	Subbasin	R^2	Accdiff
Upper Narew	Narew	0.53	43.0
	Suraż	0.62	134.0
	Fasty	0.46	30.0
	Strękowa Góra	0.53	45.0
	Osowiec	0.69	10.0
	Burzyn	0.74	3.0
	Wizna	0.72	52.0

Generally, the R^2 values fluctuated from 0.46 to 0.76 for calibration period. For many subbasins calibration results are more or less acceptable with R^2 values equal or bigger than 0.6 and not too big accumulated differences (Table 6).

In Fig. 3 graphs for some hydrological stations are presented. The visual comparison is the most important criterion and one can especially consider those parts of the hydrograph that are most essentials for the current application.

The calibration results are not entirely satisfactory and the model has not always succeeded to reproduce the observed discharge. One reason is that the precipitation network is not dense enough to describe the spatial variability of rainfall. Another reason is that the discharge records not always seem to be of good quality and correction factors for ice phenomena are not known enough (see winter and spring seasons).

The HBV model for the district Upper Narew - Subbasin Narew (the Narew River)

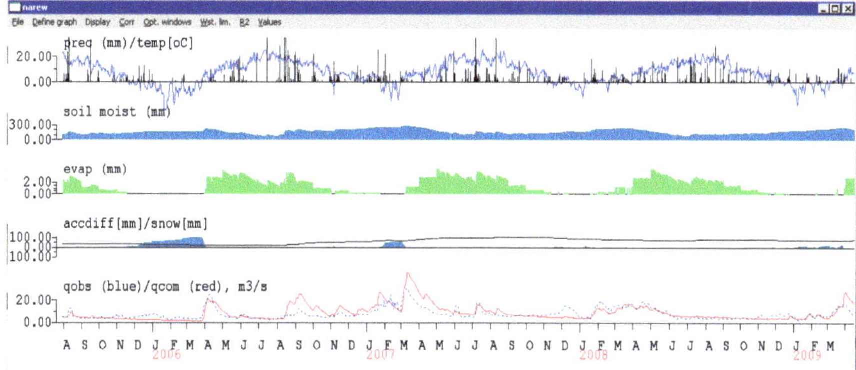

The HBV model for the district Upper Narew - Subbasin Fasty (the Supraśl River)

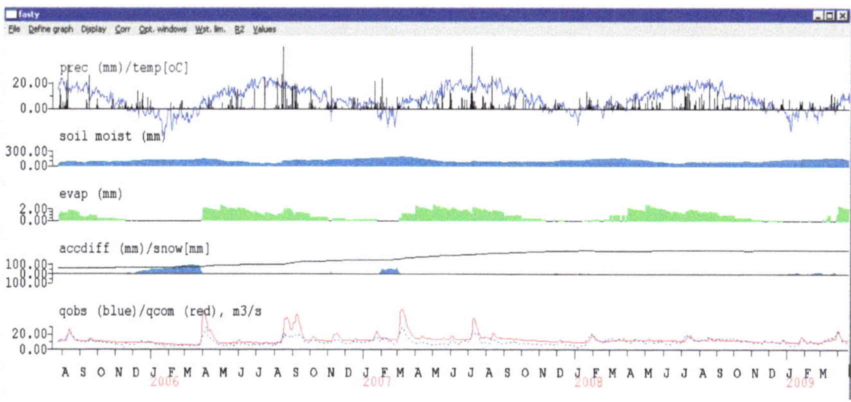

Fig. 3 The HBV model results for selected subbasins in the district Upper Narew qobs - observed (*line*), qcom - computed (*dotted line*) (continued)

For all subbasins the model underestimates the discharge during the end of the dry season in summer 2006 and the reason may be probably rain in areas without precipitation stations.

Results presented in Fig. 3 are based on operational data. It shows that significant errors occur during periods with ice phenomena. It ought to be remembered that operational data are available in every day forecasting and are verified after season based on detailed checking and analyzing of water balances. Sometimes the model result seems to be more realistic than "observed" discharge.

The HBV model for the district Upper Narew - Subbasin Strękowa Góra (the Narew River)

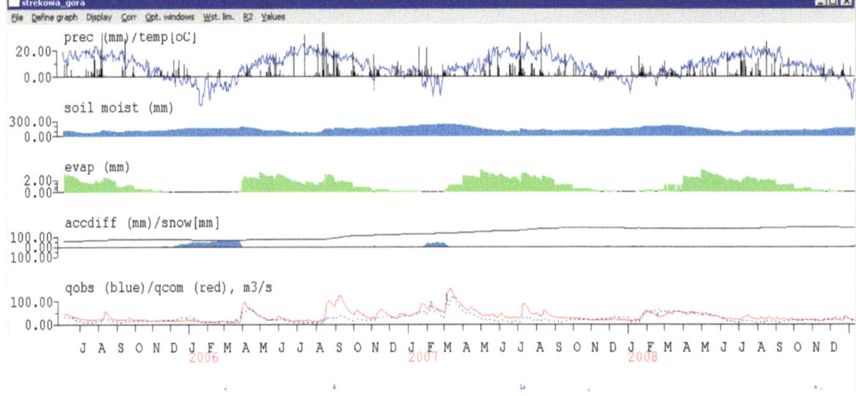

The HBV model for the district Upper Narew - Subbasin Wizna (the Narew River)

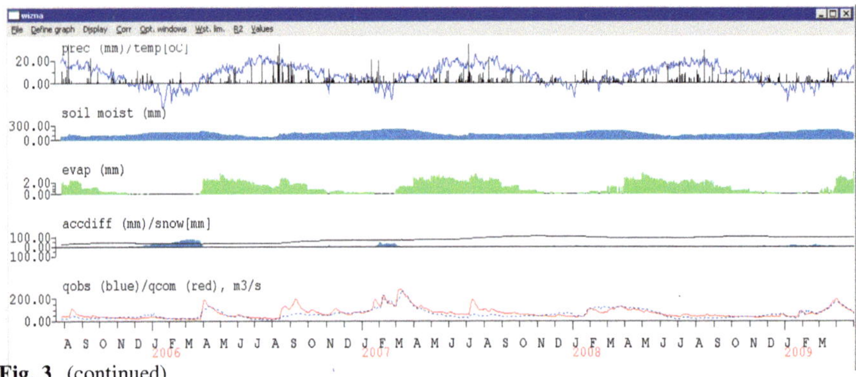

Fig. 3 (continued)

4.4 Conclusions

The main application of HBV model in Vistula River basin is to perform short-range runoff forecasts for many tributaries in the Integrated Hydrological Monitoring and Forecasting System for the Vistula River Basin (including the Upper Narew River district). The IHMS system (HBV and HD models) has been running in real-time. These forecasts are used in a flood warning system to warn people and local authorities and as an inflow to HD model prepared for the Vistula, Narew and Bug rivers.

The HBV model will be a help in the check of quality of discharge measurements and runoff data. It is also a useful tool to simulate the discharge.

The calibration process and obtained results showed that the most important factor to achieve a good model is the quality of input data, discharge, precipitation and air temperature in winter. Since the beginning of operational use of this system model HBV has been recalibrated several times based on new, longer historical time series.

References

Berström S (1976) Development and application of a conceptual model for Scandinavian catchments, SMHI, Report RHO No. 7, Norrkoping

Lindel S, Ericsson L, Mierkiewicz M, Kadłubowski A (1997) Integrated hydrological monitoring and forecasting system for Vistula River Basin. Final report, IMGW (Poland) and SMHI (Sweden)

Mierkiewicz M, Kadłubowski A, Sasim M (1999) System of hydrological protecting in the middle and lower Vistula River Basin. Wiadomości IMGW, t. XXII(XLII), z. 4 (in Polish)

Nash JE, Sutcliffe JV (1970) River flow forecasting through conceptual models. Part I. A discussion of principles. Journal of Hydrology 10:282–290

Multi-Site Calibration and Validation of the Hydrological Component of SWAT in a Large Lowland Catchment

Mikołaj Piniewski and Tomasz Okruszko

Abstract This study describes an application of a hydrological component of the catchment model, Soil and Water Assessment Tool (SWAT) in the Narew basin (ca. 28,000 km^2) situated in the north-east of Poland. The main objective was to perform a multi-site (spatially distributed) calibration and validation of SWAT using daily observed flows from 23 gauging stations as well as to assess the model's capability to perform reliable simulations at spatial scales that were smaller than those in the calibration phase. A detailed description of the model configuration for the Narew basin upstream from Zambski Kościelne gauge has been given. Building a SWAT project for a large-scale application appeared to be a demanding task, with the most critical part of preparing soil input data. Sensitivity analysis performed using a LH-OAT method indicated which parameters should be used in autocalibration. The ParaSol tool allowed to find the best parameter values from 8D parameter space in 11 calibration areas. The calibrated model generally performed well, with average Nash–Sutcliffe Efficiency for daily data equal to 0.68 for calibration period and 0.57 for validation period. SWAT correctly conserved the mass balance in different parts of the catchment as well as at the main outlet. The model results were significantly better in large basins than in small basins. Spatial validation performed at 12 independent catchments ranging in size from 355 to 1,657 km^2 revealed that adapted SWAT model should rather not be used in the Narew basin catchments smaller than ca. 600 km^2. It is believed that ensuring reliability of SWAT results at smaller spatial scales, which would be of interest to decision-makers, would require providing better input data and in particular using significantly more precipitation stations.

M. Piniewski (✉) · T. Okruszko
Department of Hydraulic Engineering, Warsaw University of Life Sciences, Nowoursynowska Str. 159, 02-787 Warszawa, Poland
e-mail: mpiniewski@levis.sggw.pl

D. Mirosław-Świątek and T. Okruszko (eds.), *Modelling of Hydrological Processes in the Narew Catchment,* Geoplanet: Earth and Planetary Sciences, DOI: 10.1007/978-3-642-19059-9_2, © Springer-Verlag Berlin Heidelberg 2011

1 Introduction

Application of computer models of catchment hydrology has become a fundamental area of research in hydrology for many years already. The term "catchment modelling" refers here to modelling spatially distributed hydrological processes over the entire catchment, which is distinct from lumped-conceptual modelling, which does not deal with spatial distribution of parameters.[1] Catchment models have always some semi-physical or physical representation of the runoff generation processes. Another feature differentiating catchment models is discretisation strategy, strongly related to their computational requirements: from fully distributed, grid-element-based models, such as Système Hydrologique Europèen (Abbott et al. 1986) and its successors such as SHETRAN (Bathurst et al. 1995) and MIKE SHE (Refsgaard and Storm 1995) to semi-distributed models built on the concept of hydrological similarity, such as TOPMODEL (Beven and Kirkby 1979) or SWAT (Arnold et al. 1998; Neitsch et al. 2005). The latter, Soil and Water Assessment Tool, is further investigated in this paper.

To the authors' current knowledge, no research results showing successful SWAT application for a large river basin in Poland have been published so far. One of the first Polish studies on SWAT dealt with creation of the tool supporting catchment management based on SWAT for the purposes of the pilot implementation of the Water Framework Directive in the Zgłowiączka catchment (Śmietanka et al. 2009). This was a small-scale application (study area of ca. 125 km^2), in which no calibration results were reported. In another study of Ulańczyk (2010), SWAT was applied in the Kłodnica catchment (437 km^2) in order to identify the causes of observed changes in water quality. Calibration and validation of the hydrological component were done for only 1 year and because of different objectives of that study, calibration method and results were not thoroughly described.

Otherwise, SWAT belongs to the most widely used catchment models in the world. Its success has many reasons, for instance its open source code, a user-friendly GIS interface or a comprehensive set of options for testing river basin management scenarios. Gassman et al. (2007) provided a comprehensive review of the whole body of SWAT applications that have been reported in the peer-reviewed literature (more than 250 articles, today this number has reached 705[2]). They also provided an extensive list of 115 papers reporting the results of hydrological calibration of SWAT, including data such as drainage areas of calibrated catchments as well as calibration and validation statistics. It was pointed out that in the great majority of these studies SWAT was calibrated and validated only at the basin's outlet. Only 5 studies reported the use of multiple gauges (from 2 to 7) to perform flow calibration and validation with SWAT:

---

[1] http://www.unesco-ihe.org/Flood-Management-Education-Platform/Flood-Modelling-for-Management2

[2] https://www.card.iastate.edu/swat_articles/ as for 21-Oct-2010.

Qi and Grunwald (2005), Cao et al. (2006), White and Chaubey (2005), Vazquez-Amábile and Engel (2005) and Santhi et al. (2001). As far as basin sizes are concerned, Gassman et al. (2007) report the studies at very widespread scales, from a plot-scale catchment of 0.015 km^2 (Chanasyk et al. 2003) to the Upper Mississippi of 491,700 km^2 (Arnold et al. 2000). The order of magnitude of areas occupied by the majority of 115 reported catchments was 10^2–10^3 km^2.

One of the major concerns that limits potential applicability of SWAT (and other similar catchment models) in Poland and perhaps explains very low interest in this model observed so far, is input data availability. Although SWAT developers (Neitsch et al. 2005) claim that this model uses readily available input data, this statement describes perhaps situation in the United States, but not necessarily in other less-developed countries, such as Poland. An access to two types of input data is particularly an issue: digital soil maps storing soil physical properties as well as climate and flow data, which are held solely by the Institute of Meteorology and Water Management in Warsaw. Our study indicates some potential paths to overcome the problems with limited data availability, which was also an important issue in the applications of SWAT in the Biobio basin in Chile by Stehr et al. (2008) and in the Pangani basin in Tanzania by Ndomba et al. (2008).

Although manual approaches are still often used for calibration of catchment models, they are tedious, time consuming, and require experienced personnel (Muleta and Nicklow 2005). The other reason for the need to use automatic approaches is that models have become more and more complex in terms of process descriptions and the number of parameters in recent years (Duan 2003). Especially in the case of a large-scale hydrological modelling [here, as in Döll et al. (2008), "large-scale" refers approximately to study areas between a few thousand km^2 and the whole globe], autocalibration seems to be a time-saving and more objective approach compared to manual calibration. On the other hand, automatic calibration can sometimes lead to improper mass balance and parameter estimates that do not conform to physical catchment characteristics and thus many studies recommend that autocalibration techniques should be combined with manual calibration to get parameter values representative of catchment conditions (Green and van Griensven 2008; van Liew et al. 2005). In this study manual calibration also serves as a supporting tool for autocalibration.

In this study the hydrological component of the SWAT model was applied to a large (28,000 km^2) basin situated in the north-east of Poland, the Narew basin. Spatially distributed automatic calibration and validation procedure was performed in order to account for hydrological differences across the basin. Santhi et al. (2008) pointed out that a traditional approach of calibrating the model only at the catchment outlet cannot account for this kind of variations, whereas the objective of this study was not only to provide reliable flow simulations in the main outlet but also in the upstream sub-basins. This could be achieved only through using multiple gauges in the processes of calibration and validation. In this study daily flow records from 11 gauges were used for model calibration and temporal validation. Additionally, spatial validation was performed at smaller spatial scale for 12 gauges in order to investigate what is the threshold of catchment size, above which using the developed model was expected to make sense.

The motivation for such scope of study comes from the issues of the integrated water management planning addressed in the Water Framework Directive (EU 2000). As it was noted by van Griensven et al. (2006a), using catchment models such as SWAT is the most appropriate reference framework for WFD-oriented integrated modelling. Basic units for making environmental assessments according to WFD are catchments of water bodies which in Poland have been merged into so-called "catchments of integrated water bodies" (CIWB). Mean area of CIWBs in the Narew basin upstream from Zambski Kościelne equals to 360 km^2. For the practical reasons, decision-makers do not apply models separately for single CIWBs but rather for large catchments such as the Narew. Therefore the question of reliability of SWAT results within the catchment, at spatial scale of several hundred square kilometres, after performing spatially distributed calibration, would be of interest to the decision-makers.

It should be noted that the validation of hydrological component of SWAT is a prerequisite for two other important objectives, which are ongoing, but beyond the scope of this paper: first, impact assessment of climate and land use changes on river flow of the Narew River, and second, an application of water quality component of SWAT. Thus, the work presented here attempts to set the basis for more advanced future modelling applications in the Narew basin.

2 Materials and Methods

2.1 Modelling Tool

SWAT is a public domain, river basin scale model developed to quantify the impact of land management practices in large, complex river basins (Arnold et al. 1998). SWAT2005 model version (Neitsch et al. 2005) under ArcSWAT 2.3 GIS interface (ArcGIS extension) was used in this study. SWAT is a physically based, semi-distributed model that operates on either daily or sub-daily time step. It can simulate the movement of water, sediment, nutrients, pesticides and bacteria in a catchment of interest. The river basin can be partitioned into a desired number of sub-basins based on the Digital Elevation Model (DEM), optionally a stream network layer and a threshold that defines the minimum drainage area required to form the origin of a stream. The smallest unit of discretisation is a unique combination of land use, soil and slope overlay referred to as a hydrologic response unit (HRU). Runoff is predicted separately for each HRU, then aggregated to the sub-basin level and routed through the stream network to the main outlet to obtain the total runoff for the river basin.

Key processes associated with the land phase of the hydrological cycle included in SWAT are (Neitsch et al. 2005; Arnold et al. 1998):

1. snowmelt estimated using degree-day method;
2. surface runoff estimated using the modified SCS curve number method or Green–Ampt infiltration equation;
3. percolation modelled with a layered storage routing technique;
4. lateral subsurface flow estimated using kinematic storage model;
5. groundwater flow to streams from shallow aquifers;
6. potential evapotranspiration by the Hargreaves, Priestley–Taylor or Penman–Monteith method;
7. actual evapotranspiration (separately evaporation from soil and transpiration from plants).

Main processes of the routing phase of the hydrological cycle represented in SWAT are:

1. channel routing modelled with the variable storage or the Muskingam river routing method;
2. transmission losses from streams;
3. water storage and losses from ponds and reservoirs.

2.2 Study Area

In this study SWAT model has been applied in the Narew basin situated in the north-east of Poland. The River Narew is a tributary of the Vistula and its drainage area is ca. 75,000 km^2. However, in this study the focus is on the part of the basin situated upstream of Lake Zegrzyńskie, which is a reservoir formed by a dam constructed in the 1960s. The first gauging station upstream from the lake, Zambski Kościelne, closes the area of ca. 28,000 km^2, 5% of which lies in western Belarus. Use of the name "the Narew basin" in this paper denotes the area described above and shown in Fig. 1, unless stated differently.

The flow regime of the Narew is typical for the large lowland floodplain rivers in Central Europe. The average yearly precipitation in 2001–2006 was 562 mm, 32% of this amount occurring in winter half-year. The peak flows occur during spring snowmelt periods while the low flows usually in late summer. The average flow recorded at Zambski Kościelne gauge between 2001 and 2006 was 114 m^3 s^{-1}, with the 391 m^3 s^{-1} peak flow in this period. It should be noted that the River Narew maintained its natural conditions, in comparison with the other river systems in Poland and in Western Europe. There is only one large man-made structure, Siemianówka reservoir situated in the Upper Narew built in the late 1980s. The Narew basin is the core part of the region known as "the Green Lungs of Poland". There are three national parks (ca. 750 km^2) protecting wetland and forest ecosystems and a number of Natura 2000 Special Protection Areas occupying ca. 27% of the Polish part of the basin. For all these reasons the Narew basin has been an interesting area for hydrological and ecological modelling, a study of Giełczewski (2003) being a good example of this.

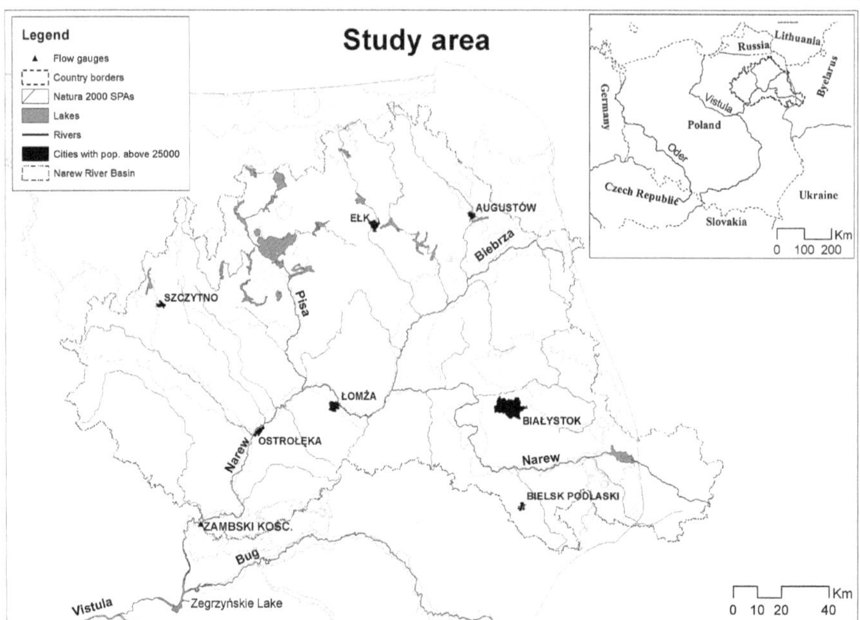

Fig. 1 Location of the study area (*SPAs* special protection areas)

2.3 Model Configuration

The main GIS and monitoring datasets used to set up SWAT project are presented in Table 1. DEM and the stream network layer were used in ArcSWAT to partition the Narew basin into 151 sub-basins (Fig. 2). An average sub-basin area equals to 182 km^2 and an average length of a reach in a sub-basin has 24 km.

The land use map and the soil map were solely used to divide each sub-basin into HRUs, whose total number in the catchment equalled to 1,131. Due to the catchment's rather flat topography, single slope option was chosen in the process of creation of HRUs. Land use classes had to be reclassified from the CLC 2000 land use classification into the SWAT classification. Three major land use classes were: arable land (44%), evergreen forests (22%) and grasslands (17%). A soil map was prepared in a stepwise procedure. First, a classification of soils from the database of benchmark soil profiles was made based on two main variables: soil texture and organic matter content. Second, a polygon layer was produced from a point feature class with locations of classified benchmark soil profiles using Thiessen polygon method. Third, an overlay of this map with the map showing the extent of hydrogenic soils was done and the result of this overlay was the final map that was used as the SWAT input. In this final map 27 soil classes were distinguished. Loamy sands (26%), sandy loams (26%) and sands (23%) were 3 dominating classes, while the share of peat soils was considerably high as well (17%). Initial values of soil parameters were set in the User Soils database based on the

Table 1 GIS and monitoring datasets used to build the SWAT project

Category	Data sources	Remarks
Digital elevation model	1. Topographic maps & 2. NASA Shuttle Radar Topography Mission (Farr et al. 2007)	1. Resolution 30 m, covering ca. 80% of the area 2. Resolution 90 m, covering remaining 20%
Stream network	Digital Map of Hydrological Division of Poland (MPHP)	From the Institute of Meteorology and Water Management in Warsaw
Land use	1. CORINE Land Cover 2000 (CLC 2000) 2. Satellite Images (Google Earth)	1. From the Institute of Geodesy and Cartography, only for the area of Poland 2. For the remaining 5% of the area in Belarus
Soil	Point feature class of 3,400 benchmark topsoil profiles	From the Institute of Soil Science and Plant Cultivation in Puławy
Climate and flow	12 rain gauge stations, 6 climate gauge stations, 11 flow gauge stations: daily data for time period 2001–2008	From the Institute of Meteorology and Water Management in Warsaw

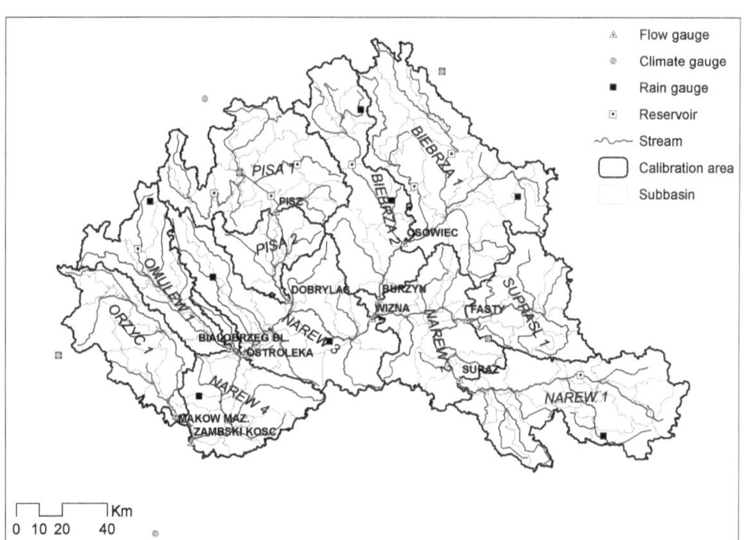

Fig. 2 Configuration of SWAT for the Narew basin

Polish literature (Zawadzki 1999; Ilnicki 2002) and pedotransfer functions (using soil texture and organic matter content as explanatory variables).

The parameters defined in the User Soils database as well as the default values of parameters from SWAT Land Cover/Plant Growth database were automatically transferred to each HRU. Therefore, all HRUs with the same soil and land use type had the same set of parameters. Further in this section it is described how some of these default parameter values were changed in the pre-calibration model setup.

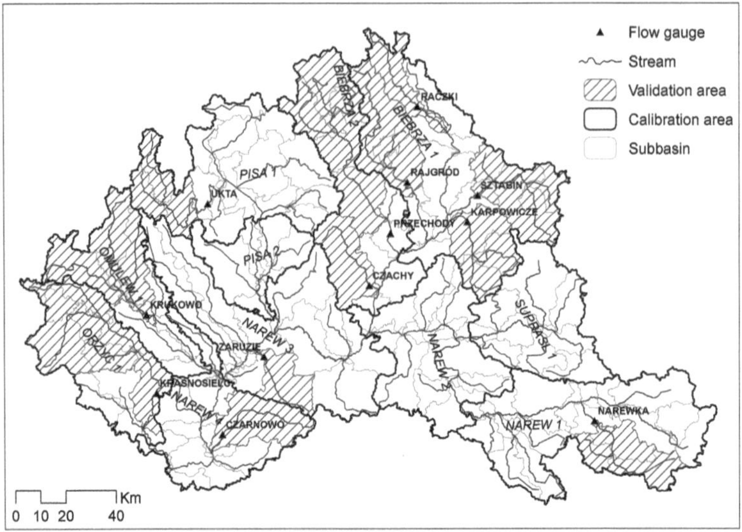

Fig. 3 Catchments selected for spatial validation (validation areas)

Daily meteorological data from 12 rain gauges and 6 climate gauges were used in this study (Fig. 2). Daily data covered the time period of hydrological (i.e., beginning on 1 November) years 2001–2008, of which the first 6 years were used for the calibration and the last 2 years for the validation. Climate data contained: minimum and maximum air temperature, relative humidity, wind speed and solar radiation (only 1 station). In order to capture spatial variability of solar radiation, gridded data from the freely available MARS-STAT dataset (van der Goot and Orlandi 2003) were additionally used. This extensive climate dataset enabled the most physically based Penman–Monteith method for modelling the evapotrans-piration, to be used. The precipitation data were interpolated from the gauging stations to sub-basins using Thiessen polygon method. This approach was an improvement over the default SWAT method of transferring data from the closest station to every sub-basin. A 2-year-long warm-up period was used in all the simulations.

Eleven gauges with daily flow records, situated on the Narew (4) and its largest tributaries: Supraśl (1), Biebrza (2), Pisa (2), Omulew (1) and Orzyc (1), were selected for the model calibration for the time period 2001–2006. Location of selected stations dictated partitioning of the whole basin into 11 subareas, to which we will further refer as "calibration areas" (Fig. 2). Temporal validation was performed for the time period 2007–2008 in all calibration areas. Spatial validation was performed for the time period 2001–2002 for 12 additional gauges situated upstream from the gauges used in the calibration (Fig. 3). The catchments delin-eated by these gauges will be further referred to as "validation areas". In several cases, when a location of gauge did not overlap with a location of sub-catchment outlet, flow time series were scaled using catchment area ratios. All selected validation areas were hydrologically independent (i.e., no catchment was downstream of another already included in the analysis), and they were much smaller than calibration areas, which was done for purpose, in order to test the ability of SWAT to produce reasonable output at small spatial scales.

Lakes situated on the main stream network can be represented in SWAT by reservoirs, for which water balance is calculated separately. There can be only one reservoir per sub-basin, therefore it was tested which lakes can potentially have significant impact on flow regime (based on such features as: lake area and volume, average outflow from the lake, etc.). Finally, only 8 lakes were selected to be modelled as reservoirs in SWAT (Fig. 2).

All GIS and climate data described above theoretically should allow a user to successfully build a SWAT project and to run simulations. In practice however, first simulation results made with default parameter values are very likely of poor quality. Some of the most important pre-calibration parameter changes that were carried out are listed below:

1. Definition of basic management operations for agricultural HRUs; since the basin size is several orders of magnitude larger than average field size, agricultural practices were defined in a simplified manner; only three crops were distinguished: spring wheat, rye and potatoes and for each of them only "Plant" and "Harvest and Kill" operations were defined.
2. Adjustment of parameters from the Land Cover/Plant Growth database, T_BASE and T_OPT: minimum and optimal temperature for plant growth.
3. Adjustment of the Manning's roughness coefficients (CH_N2 in SWAT) according to the stream size: from 0.035 (small streams) to 0.045 (large rivers). Due to lack of necessary data characterising the channel, using a more sophisticated method of defining this parameter was not possible.
4. Adjustment of channel geometry parameters; as default, SWAT assumes polynomial relationship between channel width/depth and upstream catchment area; the corrected values were approximately three times lower than default in order to conform to real channel geometry.
5. Adjustment of reservoir parameters describing reservoir geometry (surfaces and volumes for principal and emergency spillway) and a method for simulating outflow from reservoir.

SCS curve number (CN) procedure was chosen to model surface runoff and a variable storage method was used to simulate flow routing. Initially, the default CN values remained unchanged.

2.4 SWAT Autocalibration Tools

The common problem faced by the model users is that distributed hydrological models like SWAT tend to be highly over-parameterized (Beven 2002; Schuol and Abbaspour 2006). For example, every soil layer in SWAT is described by 12 physical parameters. The efficiency of the model calibration can be enhanced thanks to the sensitivity analysis of model parameters which enables the model users to concentrate on those parameters that have the biggest impact on the modelled output (Beven 2002). This process of reduction of the number of parameters, inevitable in large-scale hydrological models, is also called parameter specification.

SWAT Autocalibration Tool incorporated in ArcSWAT, including Sensitivity Analysis tool and Autocalibration tool, were used in this study. Parameter sensitivities were estimated using Latin Hypercube-One-factor-At-a-Time (LH-OAT) technique (van Griensven et al. 2006b). The procedure of LH-OAT sampling is simple. First, the distribution of each of p parameters e_1, e_2,..., e_P is subdivided into N equal strata with a probability of occurrence equal to $1/N$. Then random values of the parameters are generated such that for each of p parameters, each interval is sampled only once and the model is run N times. In the neighbourhood of each point sampled in the LH-OAT sampling, p other combinations of parameters are determined, for which the model is also run. This part of algorithm is a type of One-factor-At-a-Time sensitivity analysis method. A partial effect of a given parameter is expressed as the ratio between the relative change in model output (either average flow or objective function such as sum of squared errors between observed and simulated flow) and the relative change of this parameter, whereas a total effect or a sensitivity index $|I|$ is calculated by averaging the partial effects for each loop (van Griensven et al. 2006b). Observed daily discharges of the Narew at Zambski Kościelne from 2001 to 2006 were used as an objective function.

A set of 15 parameters representing six process categories (groundwater, soil, runoff, evaporation, channel, and snow) was selected for the sensitivity analysis. Selection was based primarily on the results from the previous studies on sensitivity analysis in SWAT (van Griensven et al. 2006b; Tattari et al. 2009; Schmalz and Fohrer 2009; White and Chaubey 2005). In the initial trials several more parameters were tested for sensitivity but since the results proved to be not sensitive, they were not reported in this study. The lower and upper parameter bounds were chosen either as default values proposed by the model developers or were based on the available literature and the data. This step of defining parameter bounds is the most critical part of sensitivity analysis due to its strong impact on the final parameter ranking (Tattari et al. 2009; Kumar and Merwade 2009). Table 2 introduces selected parameters, their definitions, categories and parameter bounds applied. In the case of several parameters that were spatially distributed (defined on a sub-basin or HRU level), relative change was preferred to absolute values.

Automatic calibration was carried out using ParaSol tool (van Griensven and Meixner 2007) which is based on the Shuffled Complex Evolution-University of Arizona (SCE-UA) optimization algorithm (Duan et al. 1992). The ParaSol method calculates objective function (OF) based on model outputs and observation time series, minimizes the OF using the SCE-UA algorithm by searching over the

Table 2 SWAT parameters and their bounds used in sensitivity analysis

SWAT name	Definition	Category	Lower bound	Upper bound
ALPHA_BF	Baseflow alpha factor (days)	Groundwater	0	1
CANMX*	Maximum canopy storage (mm)	Runoff	−50	50
CH_K2	Effective hydraulic conductivity of channel (mm/h)	Channel	0	150
CH_N2*	Manning's "n" value for the main channel (−)	Channel	−50	50
CN2*	Initial SCS runoff curve number for moisture condition II (−)	Runoff	−25	25
EPCO	Plant uptake compensation factor (−)	Evaporation	0	1
ESCO	Soil evaporation compensation factor (−)	Evaporation	0	1
GW_DELAY	Delay time for aquifer recharge (days)	Groundwater	0	50
GWQMN	Threshold depth of water in the shallow aquifer for return flow to occur (mm)	Groundwater	0	1,000
SMTMP	Threshold temperature for snowmelt (°C)	Snow	−1	2
SOL_ALB*	Moist soil albedo (−)	Soil	−25	25
SOL_AWC*	Available water capacity of the soil layer (mm mm^{-1})	Soil	−25	25
SOL_Z*	Depth from soil surface to bottom of layer (mm)	Soil	−25	25
SURLAG	Surface runoff lag coefficient (−)	Runoff	0	10
TIMP	Snow pack lag temperature (−)	Snow	0	1

* Relative change (%)

entire defined parameter space and performs uncertainty analysis with a choice between two statistical concepts (van Griensven and Meixner 2007).

The simplest approach of a single-objective function being a sum of squared flow residuals was used in this study. A typical method for conducting spatially distributed calibration using data from many gauging stations, described, e.g., in Neitsch et al. (2002) and van Liew and Garbrecht (2003) was applied. The procedure of the calibration was stepwise. First, the ParaSol was run 6 times in all the calibration areas situated in the most upstream parts of the Narew basin, that were not hydrologically connected (*Narew1, Supraśl1, Biebrza1, Pisa1, Omulew1* and *Orzyc1*), which resulted in finding optimal parameter sets for each of these subareas. In the next step, initial values of calibration parameters were adjusted to optimal values in each upstream calibration area listed above. This was followed by running the ParaSol in those of remaining calibration areas which received inflows only from the areas mentioned above (i.e., *Biebrza2* and *Pisa2*). For example, in the case of *Biebrza2*, modelled outflow from *Biebrza1* served as an inflow and the search of the parameter space was restricted only to *Biebrza2* (in *Biebrza1* calibration parameters were "frozen"). "Freezing" the parameters in upstream calibration areas was necessary due to the nested nature of selected data collection sites. This enabled the best parameter set for *Biebrza2* to be found, a different one than for the *Biebrza1*. This procedure was repeated three more times in *Narew2, Narew3* and *Narew4*. Eleven combinations of optimal parameter values were the final result, one combination per calibration area. One of the drawbacks of the calibration approach was that the parameters from the.bsn input file (i.e., those which have a single value per catchment) could not be used due to the fact that otherwise they would have different values in each calibration area, which was impossible. Fortunately, the great majority of SWAT parameters which are reported to be sensitive (van Griensven et al. 2006b; Tattari et al. 2009; Schmalz and Fohrer 2009; White and Chaubey 2005) are defined at a sub-basin or HRU level. For further discussion on some issues related to using nested data in calibration of SWAT, refer to Migliaccio and Chaubey (2007).

3 Results and Discussion

3.1 Sensitivity Analysis

The results of sensitivity analysis can be classified after Lenhart et al. (2002) in the following order ($|I|$ stands for the sensitivity index, see Sect. 2.4):

(1) Class I (very high): $|I| > 1$;
(2) Class II (high): $0.2 \leq |I| < 1$;
(3) Class III (medium): $0.05 \leq |I| < 0.2$;
(4) Class IV (low): $0 \leq |I| < 0.05$.

Table 3 illustrates results of sensitivity analysis for two criteria: Sum of Squared Errors (SSE) and Average Flow (AF). The first criterion estimates the effect of the model parameters on model performance, while the second evaluates directly the model output (total mass). The results indicate that the first effect is overall larger than the second one. There are three parameters belonging to the class I for both criteria: ESCO, CN2 and GWQMN (see Table 2 for the parameter definitions). In general, parameters representing all categories have at least medium sensitivities. The results show that surface runoff and groundwater related parameters have the highest sensitivities, which suggests that the processes of surface and subsurface runoff generation play the key role in the Narew basin. It is interesting to note that parameters controlling snow melt are apparently less sensitive than the others.

It was already mentioned that parameter ranges used in the sensitivity analysis have large influence on the final ranking and this is the reason why it is difficult to make a direct comparison of presented results with the results from the other studies. The overall importance of CN2 was reported by van Griensven et al. (2006b) for two catchments in Ohio and Texas. Tattari et al. (2009) and Schmalz and Fohrer (2009) reported the largest sensitivity of GWQMN parameter in a Finnish and German catchments respectively. White and Chaubey (2005) indicated that ESCO parameter controlling soil evaporation, was the most sensitive in the catchment in Arkansas. Other parameters from Classes I and II listed in Table 3 can also be found among the sensitive parameters in at least one of the mentioned studies. The order of parameters is different in each case, but the question remains open, whether it is because of the catchment features or parameter bounds.

Table 3 Sensitivity rankings for sum of squared errors (SSE) and average flow (AF) criteria

Class	Ranking SSE	$|I|$
Cl. I	ESCO	1.42
	CN2	1.29
	ALPHA_BF	1.26
	CH_K2	1.10
	GWQMN	1.09
Cl. II	GW_DELAY	0.95
	SOL_Z	0.44
	SOL_AWC	0.44
Cl. III	TIMP	0.18
	EPCO	0.11
	CH_N2	0.09
	SURLAG	0.06
	CANMX	0.05
Cl. IV	SMTMP	0.04
	SOL_ALB	0.01
Class	Ranking AF	$|I|$
Cl. I	CN2	1.96
	GWQMN	1.47
	ESCO	1.32
Cl. II	ALPHA_BF	0.79
	SOL_AWC	0.60
	SOL_Z	0.55
Cl. III	CH_K2	0.14
	EPCO	0.13
	CANMX	0.06
	TIMP	0.06
Cl. IV	GW_DELAY	0.03
	SMTMP	0.03
	CH_N2	0.01
	SOL_ALB	0.01
	SURLAG	0.00

3.2 Model Calibration

3.2.1 Manual Calibration

Before performing the autocalibration in a sequential procedure described earlier, SWAT was manually calibrated in order to better understand the meaning and relevance of various parameters as well as to gain control over mean annual water balance and general seasonality patterns, as suggested by Santhi et al. (2001). Mean water balance components were first estimated from the observed time series available for the time period 2001–2008. Mean runoff coefficients (defined as total runoff to precipitation ratio) ranged from 20 to 35% across the river basin. Second, the baseflow filter program described by Arnold and Allen (1999), was used in order to estimate the contribution of groundwater recharge in total water yield as well as flow recession constant (ALPHA_BF). Estimated baseflow to total runoff ratio ranged from 70 to 90% for different sub-basins. SWAT initially highly overpredicted both total water yield and especially surface runoff, therefore runoff curve numbers (CN2) were decreased to partially overcome this problem. This led to a better agreement of the modelled and observed runoff coefficients and the baseflow to total runoff ratios. Estimated ALPHA_BF values ranged from 0.01 to 0.04 and these numbers were then used as initial values in autocalibration. Next, to better simulate the timing of peaks, the default values of SURLAG and GW_DELAY, which control respectively: the lag in release of surface runoff and the lag between the time that water exits the vadose zone and enters the shallow aquifer, were decreased. Finally, it was observed that decrease of snow melt temperature (SMTMP) from the default 0.5 to 0 °C and allowing for the snow melt factors (SMFMN and SMFMX) seasonal variability, led to better simulation results in winter.

3.2.2 Autocalibration

In the next stage, the following eight parameters were selected for autocalibration: ALPHA_BF, CH_N2, CN2, EPCO, ESCO, GW_DELAY, GWQMN and SOL_AWC (see Table 2 for definitions). Such choice was a trade-off between:

1. an attempt to take into account parameters that represent different hydrological processes and are relevant for the study area (e.g., CH_K2 was considered irrelevant despite its high sensitivity);
2. time constraint on the whole process of autocalibration.

The last factor was caused by the fact that performing a single autocalibration run in 8D parameter space took several thousand single model runs, which lasted from 6 to 10 days of computations on an Intel Core2 Duo 3 GHz processor under Windows XP OS. Taking more than eight parameters would increase simulation time to the unacceptable level. This issue is addressed in the conclusions in more detail.

Parameter bounds used for autocalibration were often different than those used for sensitivity analysis: larger for SOL_AWC, the same for CH_N2, EPCO and ESCO and smaller for the rest of parameters. SOL_AWC was allowed to vary in a wider range because in a testing phase it was observed that this parameter is strongly underestimated. Narrowing of the allowed range for some parameters was done because too large ranges were not physically justified.

Table 4 summarises the results of the series of autocalibration runs from 10 out of 11 calibration areas. Results from *Supraśl1* calibration area did not meet the satisfactory threshold for goodness-of-fit measures, therefore they were not included in the table. Thresholds were met after changing the calibration scheme:

Table 4 Summary of parameters estimated in autocalibration

Cal. parameters	ALPHA_BF	CH_N2	CN2	EPCO	ESCO	GW_DELAY	GWQMN	SOL_AWC
Parameter update method[a]	1	3 (%)	3 (%)	1	1	1	1	3 (%)
Cal. area	Best values							
Narew1	0.022	−19.2	9.9	1.00	0.80	1.0	72	40
Biebrza1	0.017	28.8	−0.3	0.38	0.88	4.6	55	39
Pisa1	0.009	50.0	−10.0	1.00	0.97	10.0	73	40
Omulew1	0.009	7.9	0.9	0.00	0.91	2.1	43	40
Orzyc1	0.010	44.5	9.9	0.04	0.70	1.0	37	40
Biebrza2	0.030	−12.2	6.3	0.98	0.90	5.9	71	40
Pisa2	0.005	−0.7	−9.4	0.98	0.88	10.0	17	−21
Narew2	0.006	47.7	−9.6	0.96	0.99	2.3	67	38
Narew3	0.039	−36.5	2.9	1.00	0.91	10.0	39	35
Narew4	0.005	−30.2	1.3	0.46	1.00	2.3	40	40
Lower bound	0.005[b]	−50	−10	0	0	1	0	−40
Upper bound	0.04[b]	50	10	1	1	10	100	40
Average	0.015	8.0	0.2	0.68	0.89	4.9	52	33
Std. dev.	0.012	33.0	7.6	0.42	0.09	3.8	19.2	19.1

[a] *1*, replacement by a new value; *3*, relative change in percent
[b] For ALPHA_BF lower and upper bounds varied slightly across calibration areas: lower bound from 0.005 to 0.01 and upper bound from 0.02 to 0.04

observed data from one more gauge were used and parameter ranges were slightly widened. In Table 4 calibration areas are grouped into five classes corresponding to sequential steps of autocalibration process, as described in Sect. 2.4. The lower and upper bounds of parameter variation were the same in each calibration area, apart from the ALPHA_BF. For parameters: CH_N2, CN2 and SOL_AWC the numbers in the table are percent changes from the initial values, whereas for the others these are the final best parameters. The range of initial values for CH_N2 was from 0.035 to 0.045 (depending on the stream size), for CN2 from 36 to 93 (depending on the soil and land use type) and for SOL_AWC from 0.12 to 0.5 (depending on the soil type).

It can be observed that in each calibration area different parameter sets were found. ESCO was the only parameter, whose values in different calibration areas were similar. This parameter was earlier found to be very sensitive (Table 3). Physically, it represents the effects of capillary action in soils. As the value for ESCO approaches 1, which was the case in most of calibration areas, the model is able to extract less evaporative demand from the lower layers. There were two calibration areas (*Narew1* and *Orzyc1*) for which calibrated CN2 values were close to +10% (increase to initial values) and three for which they were close to −10% (*Pisa1*, *Pisa2* and *Narew2*). An increase of CN2 implies an increase of surface runoff and a decrease in baseflow. Regional variability in CN2 is in accordance with the estimates obtained from the baseflow filter program, which indicate that surface runoff constitutes as much as 26–30% of total runoff in *Narew1* and *Orzyc1* and as low as 10–13% in *Pisa1*, *Pisa2* and *Narew2*. The results for SOL_AWC suggest that a priori estimates of this soil parameter found in Zawadzki (1999) and Ilnicki (2002) should be increased by ca. 40% to obtain closer fit between the observed and simulated values.

3.3 Model Evaluation

3.3.1 Goodness-of-Fit Measures

Basic modelled and observed flow statistics and the most widely used goodness-of-fit measure, Nash–Sutcliffe Efficiency (NSE, Nash and Sutcliffe 1970), are given in Table 5 for all the calibration areas. A general acceptable threshold for NSE is reported to be 0.5 for the monthly data (Moriasi et al. 2007). In this study a threshold of 0.5 was used for the daily data which is certainly more stringent. Minimising the objective function in autocalibration is equivalent to maximising NSE, but another important aspect of model evaluation is the control of the mass balance. In Moriasi et al. (2007) a maximum 25% bias in mean flow was considered acceptable. Table 4 presents also standard deviations of daily flows which depict agreement in flow variability between simulated and observed data and RMSE, Root Mean Squared Error, which estimates error in absolute numbers.

Results for calibration period were overall positive, with the average, minimum and maximum NSE equal to 0.68, 0.55 and 0.82, respectively. Mass balance was also well conserved, with maximum underestimation of 16.2% for *Biebrza2*. Variability represented by standard deviation was modelled appropriately, with maximum underestimation of variability equal to 26.9% also for *Biebrza2*. Results for validation period were overall a little worse than for calibration, with the average, minimum and maximum NSE equal to 0.57, −0.09 and 0.84, respectively. Model performance in *Narew1* was clearly bad during this period, which will be explained later. The model performance in *Orzyc1* and *Pisa1* was also unsatisfactory, although to a smaller extent than in *Narew1*. Mass balance was well conserved in validation period, similarly to calibration period. The biggest underestimation of average flow ranging from 15.2 to 16.9% occurred in *Omulew1*, *Orzyc1*, *Biebrza1* and *Biebrza2*. Flow variability was simulated well everywhere apart from *Narew1*.

It is worth mentioning that the model performance in simulating monthly flows was yet a little better than for daily flows. Monthly NSE was greater than daily NSE for each gauging station, both in calibration and validation period, on average 0.04 and in the best case 0.15. It is equally important to note that in the pre-calibration runs SWAT performed in a significantly worse manner, with most frequently negative NSE for the majority of gauging stations under consideration.

Table 5 Evaluation of SWAT during calibration and temporal validation phases

Cal. area	Calibration 2001–2006						Validation 2007–2008					
	Q_{obs} (m^3/s)		Q_{sim} (m^3/s)		RMSE (m^3/s)	NSE (–)	Q_{obs} (m^3/s)		Q_{sim} (m^3/s)		RMSE (m^3/s)	NSE (–)
	Av	SD	Av	SD			Av	SD	Av	SD		
Narew1	11.4	9.43	10.0	8.74	5.37	0.67	15.7	8.80	15.3	14.9	9.19	−0.09
Supraśl1	7.80	4.22	8.14	3.99	2.66	0.60	9.82	4.56	9.87	3.50	2.80	0.62
Biebrza1	18.6	14.8	16.4	12.4	8.28	0.69	24.2	15.4	20.3	14.7	7.90	0.74
Pisa1	14.9	6.22	15.2	6.06	3.64	0.66	20.4	5.69	22.6	5.90	4.44	0.39
Omulew1	8.41	5.01	8.38	5.00	3.23	0.59	8.67	5.00	7.35	5.08	3.27	0.57
Orzyc1	6.89	6.90	6.78	5.48	4.65	0.55	6.67	5.63	5.58	5.24	4.83	0.26
Narew2	56.2	39.4	54.1	36.0	18.3	0.78	72.4	43.2	68.9	45.2	17.3	0.84
Biebrza2	31.8	29.7	26.6	21.7	18.5	0.61	38.6	24.4	32.1	23.3	11.9	0.76
Pisa2	19.2	7.34	19.5	7.25	3.91	0.72	24.9	7.44	27.5	7.17	5.17	0.52
Narew3	89.5	55.5	85.4	54.4	23.9	0.81	112	57.2	107	64.3	22.5	0.84
Narew4	114	68.1	111	67.1	29.1	0.82	137	70.9	128	77.1	29.8	0.82
Average						0.68						0.57

Q_{obs}/Q_{sim}, observed/simulated daily flow; *Av*, average; SD, standard deviation; RMSE, root mean squared error; NSE, Nash–Sutcliffe efficiency

Table 6 Evaluation of SWAT during spatial validation phase

Val. area		Cal. area	Area (km^2)	Spatial validation 2001–2002					
				Q_{obs} (m^3/s)		Q_{sim} (m^3/s)		RMSE (m^3/s)	NSE (–)
River	Gauge			Av	SD	Av	SD		
Narewka	Narewka	*Narew1*	617	2.05	2.30	2.32	2.75	1.86	0.35
Biebrza	Sztabin	*Biebrza1*	853	4.03	3.67	2.41	2.46	2.90	0.37
Rospuda	Raczki	*Biebrza1*	383	3.36	2.79	3.61	2.33	2.08	0.44
Brzozówka	Karpowicze	*Biebrza1*	712	2.84	3.78	2.42	2.33	3.19	0.29
Jegrznia	Rajgród	*Biebrza1*	835	3.49	3.60	3.99	3.27	2.07	0.67
Ełk	Przechody	*Biebrza2*	1,657	10.59	8.55	7.56	7.83	5.26	0.62
Wissa	Czachy	*Biebrza2*	536	2.47	2.72	2.51	2.84	2.32	0.27
Krutynia	Ukta	*Pisa1*	568	1.34	1.48	1.86	2.15	1.79	−0.47
Ruż	Zaruzie	*Narew3*	355	3.47	1.16	4.44	1.45	1.29	−0.25
Omulew	Krukowo	*Omulew1*	1,121	7.28	3.87	8.49	4.21	2.36	0.63
Orz	Czarnowo	*Narew4*	594	1.89	2.97	4.21	1.39	3.35	−0.27
Orzyc	Krasnosielc	*Orzyc1*	1,507	7.58	7.15	6.92	4.50	5.28	0.45
	Average		812						0.24

Abbreviations as in Table 5. Catchment areas as modelled in SWAT may differ from actual

The results of spatial validation of SWAT for the time period 2001–2002 are presented in Table 6. Location of validation areas was shown in Fig. 3. Only 2-year-long simulation period does not allow to draw specific conclusions but it seems to be clear that at spatial scales of 300–1,700 km^2 the ability of SWAT to simulate flows in the Narew basin properly is significantly lower than at the larger scales used in the calibration phase. The simulation results are particularly poor for three sites: Ukta, Zaruzie and Czarnowo. The possible reasons are: (1) precipitation and climate data used in the model were not representative for these catchments; (2) parameterisation of these catchments was improper. It is worth noting that optimal parameter combinations found by the ParaSol were unique in the whole calibration areas and since these areas were not spatially homogeneous it is likely that some sub-basins were poorly modelled.

Figure 4 illustrates the relationships between catchment areas upstream from the gauging stations and model performance expressed by daily NSE for calibration, temporal and spatial validation phases. Strong exponential relationship can be observed for calibration data and less strong, due to more scattered NSE values for smaller sub-catchments, for validation data. The largest differences can be observed between three biggest catchments (the Narew upstream from Wizna, Ostrołęka and Zambski Kościelne), where SWAT performs very well and the rest of catchments, where its behaviour can be both good and bad. We suggest a few possible reasons, which could explain the relationships illustrated in Fig. 4:

1. NSE is known to be biased towards high flows (Moriasi et al. 2007). In small upstream basins the response to local rainfall is very quick and their flow curves

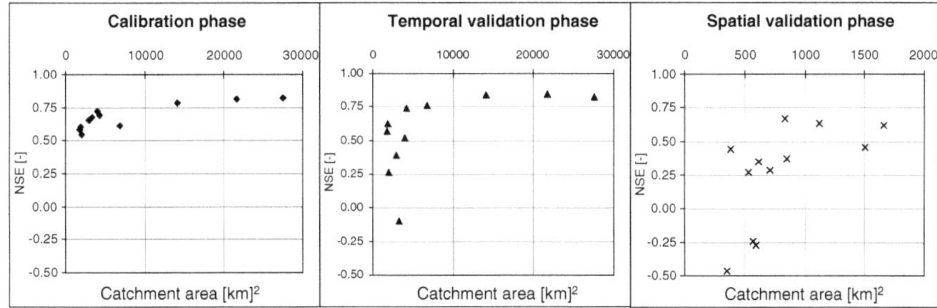

Fig. 4 Relationships between catchment area and NSE for gauges used in calibration and validation

are very flashy, whereas in larger downstream basins flows result from the sum of many small streams and in the consequence the hydrographs are much smoother. This explains that it is generally more difficult for a small catchment to have high NSE.

2. The smaller the catchment area, the more error in the input data such as precipitation, land use and soils, or in other words, spatial heterogeneity gets lower.

3. As reported by Qi and Grunwald (2005), over- and underestimations of river flow in the small upstream basins may average out in the downstream basin. A similar phenomenon was observed also in this study, e.g., in temporal validation phase, the flow of the Narew at Ostrołęka was 4% underestimated, whereas in the catchments situated upstream the direction of bias was different and its magnitude was larger: flow of the Biebrza at Burzyn was 17% underestimated, whereas flow of the Pisa at Pisz was 11% overestimated.

4. The results are influenced by the model calibration design.

3.3.2 Visual Inspection

Statistical analysis of the model performance should always be made in parallel with graphical analysis of time series plots. Figure 5 illustrates measured and simulated daily hydrographs for calibration and validation periods, for seven out of eleven stations used in the calibration. Visual inspection can bring several conclusions on the issues of model performance:

1. SWAT tends to underpredict largest spring snowmelt floods, as those that happened in 2002 and 2005. This is particularly the case at Suraż, Burzyn, Białobrzeg Bliższy and Maków Mazowiecki gauging stations.

2. SWAT tends to overpredict some of the freshets caused by precipitation, as those that happened in the summers of 2004 and 2006. This is the case at Burzyn, Dobrylas, Maków Mazowiecki and Zambski Kościelne (2004 and 2006), Suraż and Białobrzeg Bliższy (2006) and Fasty (2004).

3. SWAT tends to underpredict some of the low flow periods, in particular those that happened in winters 2001 and 2006. This is the case at every station,

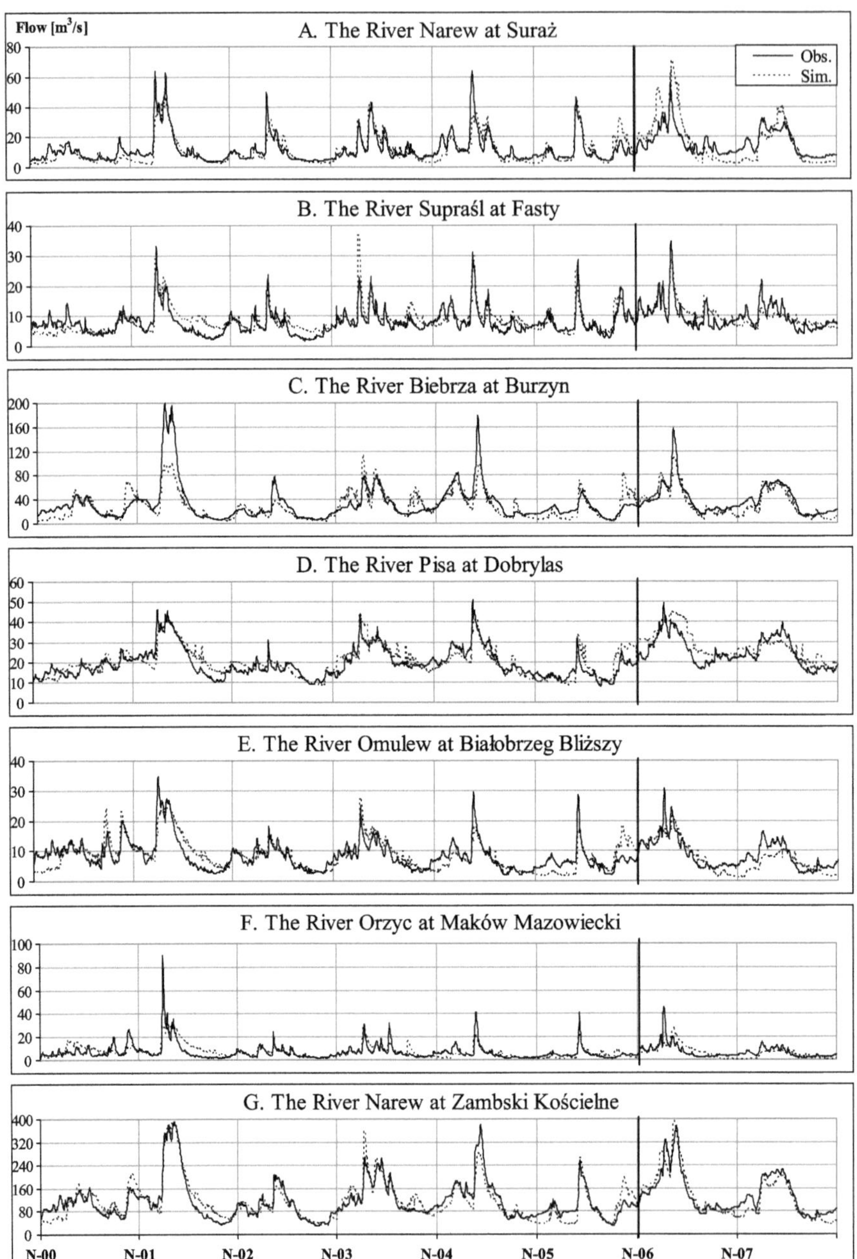

Fig. 5 Calibration and validation plots for selected stations (*black vertical line* separates calibration and validation periods, "N" stands for November)

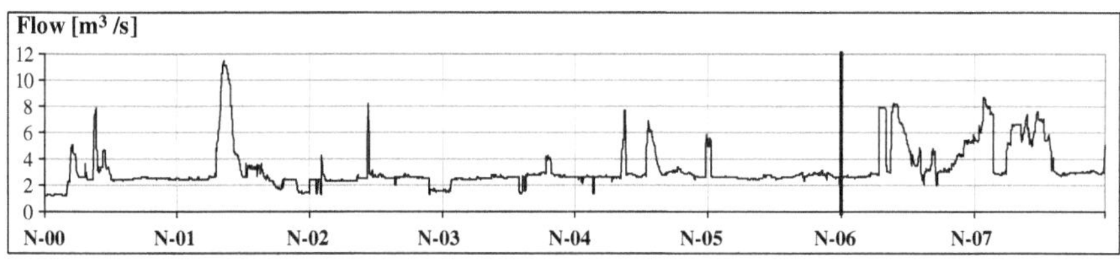

Fig. 6 Flow release from Siemianówka reservoir in 2001–2008 (*black vertical line* separates calibration and validation periods, "*N*" stands for November)

although at Suraż, Burzyn, Białobrzeg Bliższy and Maków Mazowiecki to the largest extent; it is very likely caused by the fact that some HRUs during very dry conditions stop to generate baseflow. It is difficult to say why this happens, but it may be linked to the improper values of certain groundwater parameters or data deficiencies (e.g., not representative precipitation);

4. Apparently bad model behaviour at Suraż gauging station during the validation can be explained to a large extent by the differences in flow releases from the Siemianówka reservoir between 2001–2006 and 2007–2008 (Fig. 6); it can be observed that the reservoir release policy in the second period was altered. For instance the flow release rate was gradually increasing from September to December 2007 and apparently in this period SWAT was strongly underestimating flow at Suraż (Fig. 5a). Similar problem was reported by Schuol and Abbaspour (2006) in their case study of SWAT calibration for the River Niger.

The positive points coming from the graphical analysis of calibration plots are the following:

1. The timing of peaks is overall correct (with a tolerance limit of several days);
2. Despite some cases of underprediction and overprediction of peak flows, there is a considerable amount of floods during the whole simulation period in all analysed catchments that are well modelled.

3.3.3 General Discussion

Even though SWAT is reported by its developers (Neitsch et al. 2005) to use readily available data, the experience from this study suggests that in the case of large catchments situated in the countries with a limited availability of free of charge data, building SWAT model configuration may be a task in itself (in terms of devoted time and/or money). The most critical part is the preparation of the soil map and estimation of the soil parameters. For the latter, performing sensitivity analysis can significantly help to identify to which parameters more attention should be paid and which can be ignored due to their insensitivity. It was also observed that a lot of SWAT "physical" parameters (e.g., channel geometry parameters, plant database parameters) had their default values inappropriate for the case study. This limitation may be due to the fact that these values were based on the research and applications carried out mainly in the

United States, which do not necessarily apply everywhere (e.g., in Poland). Finally, it should be noted that in our setup of SWAT for the Narew basin the management practices were treated in a very simplified manner mainly due to difficult access to appropriate data sources at such large scale. It is believed that including information on such practices as: (1) open ditch drainage on grasslands and tile drainage on arable land; (2) fertilisation of all agricultural land, should improve the results of hydro-logical component calibration.

Autocalibration has several advantages over the manual calibration, however one of the most problematic issues is defining proper parameter bounds for those parameters which are to be used in the optimisation stage. These ranges strongly affect autocali-bration results. In numerous SWAT applications authors often report parameter ranges used in autocalibration and quite frequently these values are different for the same parameters while rather seldom is the usage of a particular range explained. For some parameters, such as soil available water capacity (SOL_AWC), it is possible to use values provided in extensive literature or even measure them directly in the field, however it has to be remembered that small-scale values than can be measured and the effective values required at the model scale are different quantities, even if they had been given the same name by the hydrologists (Beven 2002). The physical relevance of the optimal parameters found in this study is quite limited, which is usually the case in large-scale hydrological modelling. Therefore it should not be considered surprising that, e.g., the optimal values for SOL_AWC were in most cases 40% higher than values reported in the Polish literature (Zawadzki 1999; Ilnicki 2002).

Time limitation has to be kept in mind, while attempting to use SWAT Aut-ocalibration Tools in multi-site calibration. The experience arising from this study indicates that the amount of time that is needed to finish autocalibration process is proportional to:

1. the number of years with observed data used in autocalibration;
2. the number of stations used in autocalibration;
3. the number of parameters used in autocalibration;
4. model configuration aspects, e.g., the number of sub-basins and HRUs;
5. several SCE-UA input parameters.

In the described case study, with 8 years of daily data from 11 stations, 8D parameter space, 151 sub-basins and 1,131 HRUs and most of SCE-UA parameters kept as default, the total amount of time used for autocalibration yielded approximately 85 days of computer work (pure model runs) on an Intel Core2 Duo 3 GHz processor under Windows XP OS.

3.3.4 SWAT Applicability in Small Catchments

The analysis of Figs. 4 and 5 and Tables 5 and 6 gives an insight on the appli-cability of the adapted SWAT model to simulate flows in the catchments belonging to the Narew basin, of smaller size than the mean size of the calibration areas. Basin size is certainly not the only factor explaining the model performance,

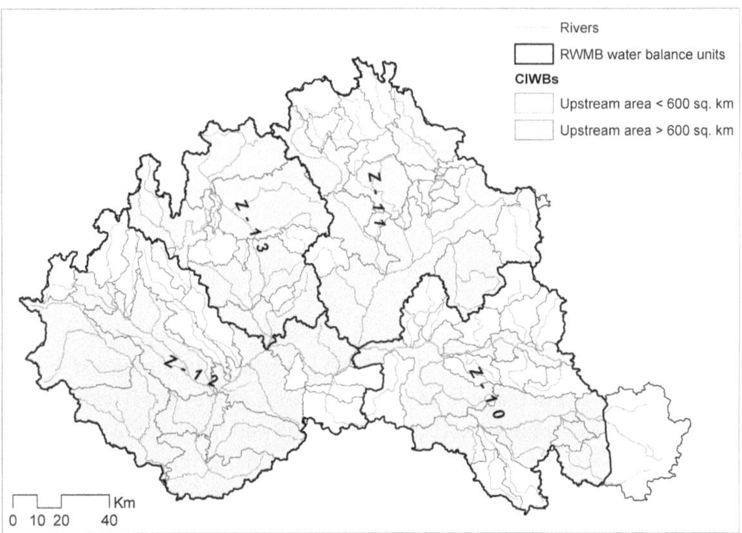

Fig. 7 Potential applicability of SWAT in units used by Polish water administration (*RWMB* regional water management board; *Z-10–Z-13*, names of the RWMB water balance units; *CIWBs*, catchments of integrated water bodies)

however, this study has shown that it is an important one. Firstly, it should be noted that for 5 catchments that were smaller than 600 km^2 (gauges: Raczki, Czachy, Ukta, Zaruzie and Czarnowo) mean NSE equalled to −0.05 and the mean percent bias was 40% (as the absolute value). This shows that the model results were very poor for these areas and suggests that for the catchments in the Narew basin smaller than ca. 600 km^2 the model would likely give unreliable estimates. For another 7 catchments ranging in size from 600 to 1,700 km^2 (gauges: Narewka, Sztabin, Karpowicze, Rajgród, Przechody, Krukowo and Krasnosielc) the mean NSE equalled to 0.48 and mean percent bias was 19%. This indicates that SWAT simulations at these spatial scales were acceptable, or almost acceptable.

One could argue that the temporal validation in the *Narew1* calibration area (Suraż gauge) indicates that SWAT performed badly in the catchment occupying over 3,000 km^2 (see Fig. 5a; Table 5) which would contradict the previous statements. However, as it was discussed in the previous section, the poor performance can be explained by the operation of Siemianówka reservoir, which was not adequately represented in the model.

As discussed in the introduction, the motivation for this study has come from the need of making WFD-related assessments in the CIWBs. Figure 7 illustrates the map of CIWBs in the Polish part of the Narew basin. CIWBs whose upstream catchment area is greater than 600 km^2 were marked on the map. This map gives an approximate indication which CIWBs flow simulations are likely to be acceptable. Selected catchments constitute 32 out of 74 CIWBs and occupy approximately 71% of the Polish part of the Narew basin. The authors suppose that the use of adapted

SWAT model for simulating flow in the remaining CIWBs would involve a risk of making high prediction errors. Figure 7 also illustrates the water balance units used by the Regional Water Management Board in Warsaw, which are one level higher in the spatial division of the Polish waters. There are 4 units of this type in the Narew basin and the results presented above clearly confirm that adapted SWAT model could be operationally used at that spatial scale.

4 Conclusions

Model evaluation results indicate that SWAT adequately simulated hydrographs and conserved the mass balance in an acceptable range in all calibration areas. Spatial validation allowed to estimate the threshold for catchment areas below which the model should rather not be applied (in the current configuration of SWAT for the Narew basin), which was approximately 600 km^2. This allowed to create a map of the potential applicability of SWAT in the CIWBs (Fig. 7). This map included all the Polish part of the Narew and Narewka, Pisa, Orzyc, Brzozówka, Orz and Krutynia, large parts of the Biebrza, Supraśl, Omulew, Ełk and Netta and the lower part of the Jegrznia. It is believed that the applicability of SWAT in all other CIWBs, which are mainly small tributaries of the Narew and Supraśl, would require better input data, in particular:

1. more precipitation stations and more flow gauging stations for spatial validation in small catchments;
2. using a real soil map instead of transferring soil information from points to polygons;
3. including agricultural management practices such as drainage and fertilisation;
4. daily reservoir/lake releases, especially from the Siemianówka reservoir.

It is difficult to judge which of these points should bring the biggest improvement to SWAT simulations. In our opinion, the lack of sufficient number of precipitation stations is the largest drawback in the current approach. It would be of interest to decision-makers to know how many stations need to be used to ensure satisfactory performance of SWAT at the level of CIWBs and such study is planned to be carried out in the future.

SWAT Autocalibration Tools used in this study proved to be successful in identifying sensitive parameters and fine-tuning the model using data from multiple stations. The authors recommend using these tools or similar ones, such as more recently created SWAT-CUP (SWAT Calibration and Uncertainty Programs; Abbaspour 2007) in each case study with the SWAT model, because the optimal parameter values found in one study area cannot be easily transferred to other study areas (Beven 2002). For catchments with similar topography, land use, soils and climate, the values reported in Table 4 can be tested in manual calibration phase or as the initial values in autocalibration phase—they should at least allow for better results than when using the default values. It should also be noted that the calibration of a distributed catchment model does not lead to the unique solutions and completely different combinations of parameter values can lead to equally good results (equifinality, see e.g., Beven and Binley 1992). This is connected to the uncertainties associated with model parameters, however these issues have not been addressed in this paper.

The concluding remark is that the basis for future applications of SWAT model in the Narew basin, including analysing climate and land use change scenarios as well as calibration of sediment and nutrient component, has now been set. Depending on the specific type of the future model use, some modifications of its configuration or even recalibration may appear to be necessary.

Acknowledgments The authors gratefully acknowledge financial support for the project *Water Scenarios for Europe and Neighbouring States* (SCENES) from the European Commission (FP6 contract 036822). We would also like to thank two referees for their constructive comments and improvement of English and Prof. Raghavan Srinivasan from Texas A&M University for clarifying many SWAT-related issues.

References

Abbaspour KC (2007) User manual for SWAT-CUP, SWAT calibration and uncertainty analysis programs. Swiss Federal Institute of Aquatic Science and Technology, Eawag, Dübendorf. http://www.eawag.ch/forschung/siam/software/swat. Last Accessed March 2011

Abbott MB, Bathurst JC, Cunge JA, O'Connell PE, Rasmussen J (1986) An introduction to to the European Hydrological System—Système Hydrologique Europèen, SHE. 1 History and philosophy of a physically-based, distributed modelling system. J Hydrol 87:61–77

Arnold JG, Allen PM (1999) Automated methods for estimating baseflow and groundwater recharge from streamflow records. J Am Water Resour Assoc 35(2):411–424

Arnold JG, Srinavasan R, Muttiah RS, Williams JR (1998) Large area hydrologic modelling and assessment. Part 1. Model development. J Am Water Resour Assoc 34:73–89

Arnold JG, Muttiah RS, Srinivasan R, Allen PM (2000) Regional estimation of base flow and groundwater recharge in the upper Mississippi basin. J Hydrol 227(1–4):21–40

Bathurst JC, Wicks JM, O'Connell PE (1995) The SHE/SHESED basin scale water flow and sediment transport modelling system. In: Singh VP (ed) Computer models of watershed hydrology. Water Resource Publications, Highlands Ranch, pp 563–594

Beven KJ (2002) Rainfall-runoff modelling: the primer. Wiley, Chichester

Beven KJ, Binley A (1992) The future of distributed models: model calibration and uncertainty prediction. Hydrol Process 6:279–298

Beven KJ, Kirkby MJ (1979) A physically based, variable contributing area model of basin hydrology. Hydrol Sci Bull 24(1):43–69

Cao W, Bowden WB, Davie T, Fenemor A (2006) Multi-variable and multi-site calibration and validation of SWAT in a large mountainous catchment with high spatial variability. Hydrol Process 20(5):1057–1073

Chanasyk DS, Mapfumo E, Willys W (2003) Quantification and simulation of surface runoff from fescue grassland watersheds. Agric Water Manag 59(2):137–153

Döll P, Berkhoff K, Bormann H, Fohrer N, Gerten D, Hagemann S, Krol M (2008) Advances and visions in large-scale hydrological modelling: findings from the 11th workshop on large-scale hydrological modelling. Adv Geosci 18:51–61

Duan Q (2003) Global optimization for watershed model calibration. In: Duan Q, Gupta HV, Sorooshian S, Rousseau AN, Turcotte R (eds) Calibration of watershed models. American Geophysical Union, Washington, pp 89–104

Duan Q, Sorooshian S, Gupta VK (1992) Effective and efficient global optimization for conceptual rainfall-runoff models. Water Resour Res 28(4):1015–1031

EU (2000) Water Framework Directive. Council Directive 2000/6/EG, 22.12.2000

Farr TG et al (2007) The shuttle radar topography mission. Rev Geophys 45:RG2004

Gassman PW, Reyes MR, Green CH, Arnold JG (2007) The soil and water assessment tool: historical development, applications, and future research directions. Trans ASABE 50(4):1211–1250

Giełczewski M (2003) The Narew River Basin: a model for the sustainable management of agriculture, nature and water supply. Netherlands Geographical Studies 317, Utrecht

Green CH, van Griensven A (2008) Autocalibration in hydrologic modeling: using SWAT2005 in small-scale watershed. Environ Model Softw 23:422–434

Ilnicki P (ed) (2002) Peatlands and peat. Wydawnictwo Akademii Rolniczej w Poznaniu, Poznań (in Polish)

Kumar S, Merwade V (2009) Impact of watershed subdivision and soil data resolution on SWAT model calibration and parameter uncertainty. J Am Water Resour Assoc 45(5):1179–1196

Lenhart T, Eckhardt K, Fohrer N, Frede H-G (2002) Comparison of two different approaches of sensitivity analysis. Phys Chem Earth 27:645–654

Migliaccio KW, Chaubey I (2007) Comment on W. Cao, B. W. Bowden, T. Davie, and A. Fenemor. 2006. Multi-variable and multi-site calibration and validation of SWAT in a large mountainous catchment with high spatial variability. *Hydrol Process* 21(4): 3326–3328

Moriasi DN, Arnold JG, van Liew MW, Bingner RL, Harmel RD, Veith TL (2007) Model evaluation guidelines for systematic quantification of accuracy in watershed simulations. Trans ASABE 50(3):885–900

Muleta MK, Nicklow JW (2005) Sensitivity and uncertainty analysis coupled with automatic calibration for a distributed watershed model. J Hydrol 306:127–145

Nash JE, Sutcliffe JV (1970) River flow forecasting through conceptual models. Part I—a discussion of principles. J Hydrol 125:277–291

Ndomba P, Mtalo F, Killingtveit A (2008) SWAT model application in a data scarce tropical complex catchment in Tanzania. Phys Chem Earth 33:626–632

Neitsch SL, Arnold JG, Kiniry JR, Srinivasan R, Williams JR (2002) Soil and water assessment tool user's manual. Version 2000. GSWRL-BRC, Temple

Neitsch SL, Arnold JG, Kiniry JR, Williams JR (2005) Soil and water assessment tool theoretical documentation. Version 2000. GSWRL-BRC, Temple

Qi C, Grunwald S (2005) GIS-based hydrologic modeling in the Sandusky watershed using SWAT. Trans ASABE 48(1):169–180

Refsgaard J-C, Storm B (1995) MIKE SHE. In: Singh VP (ed) Computer models of watershed hydrology. Water Resource Publications, Highlands Ranch, pp 809–846

Santhi C, Arnold JG, Williams JR, Dugas WA, Srinivasan R, Hauck LM (2001) Validation of the SWAT model on a large river basin with point and nonpoint sources. J Am Water Resour Assoc 37(5):1169–1188

Santhi C, Kannan N, Arnold JG, Di Luzio M (2008) Spatial calibration and temporal validation of flow for regional scale hydrologic modelling. J Am Water Resour Assoc 44(4):829–846

Schmalz B, Fohrer N (2009) Comparing model sensitivities of different landscapes using the ecohydrological SWAT model. Adv Geosci 21:91–98

Schuol J, Abbaspour KC (2006) Calibration and uncertainty issues of a hydrological model (SWAT) applied to West Africa. Adv Geosci 9:137–143

Śmietanka M, Brzozowski J, Śliwiński D, Smarzyńska K, Miatkowski Z, Kalarus M (2009) Pilot implementation of WFD and creation of a tool for catchment management using SWAT: River Zglowiaczka Catchment Poland. Front Earth Sci 3(2):175–181

Stehr A, Debels P, Romero F, Alcayaga H (2008) Hydrological modelling with SWAT under conditions of limited data availability: evaluation of results from a Chilean case study. Hydrol Sci J 53(3):588–601

Tattari S, Koskiaho J, Bärlund I, Jaakkola E (2009) Testing a river basin model with sensitivity analysis and autocalibration for an agricultural catchment in SW Finland. Agric Food Sci 18:428–439

Ulańczyk R (2010) Application of catchment water balance model to determine reasons for change in surface water quality—case study for the Kłodnica river. In: Innovative solutions for reclamation of degraded areas. CBiDGP & IETU, Ledziny-Katowice, pp 196–204 (in Polish)

van der Goot E, Orlandi S (2003) Technical description of interpolation and processing of meteorological data in CGMS. Institute for Environment and Sustainability, Ispra (http://mars.jrc.it/mars/About-us/AGRI4CAST/Data-distribution)

van Griensven A, Meixner T (2007) A global and efficient multi-objective auto-calibration and uncertainty estimation method for water quality catchment models. J Hydroinform 09.4:277–291

van Griensven A, Breuer L, Di Luzio M, Vandenberghe V, Goethals P, Meixner T, Arnold J, Srinivasan R (2006a) Environmental and ecological hydroinformatics to support the implementation of the European Water Framework Directive for river basin management. J Hydroinform 08.4:239–252

van Griensven A, Meixner T, Grunwald S, Bishop T, Diluzio M, Srinivasan R (2006b) A global sensitivity analysis tool for the parameters of multi-variable catchment models. J Hydrol 324:10–23

van Liew MW, Garbrecht J (2003) Hydrologic simulation of the Little Washita River experimental watershed using SWAT. J Am Water Resour Assoc 39(2):413–426

van Liew MW, Arnold JG, Bosch DD (2005) Problems and potential of autocalibrating a hydrologic model. Trans ASABE 48(3):1025–1040

Vazquez-Amábile GG, Engel BA (2005) Use of SWAT to compute groundwater table depth and streamflow in the Muscatatuck River watershed. Trans ASABE 48(3):991–1003

White KL, Chaubey I (2005) Sensitivity analysis, calibration, and validations for a multisite and multivariable SWAT model. J Am Water Resour Assoc 41(5):1077–1089

Zawadzki S (ed) (1999) Soil science. PWRiL, Warszawa (in Polish)

Automatic Calibration of the WetSpa Distributed Hydrological Model for Small Lowland Catchments

Laura Porretta-Brandyk, Jarosław Chormański, Andrzej Brandyk and Tomasz Okruszko

Abstract Contemporary requirements of environment protection created the need to study the hydrological regime of those catchments in which valuable and endangered wetland habitats exist. In consequence, water management scenarios are elaborated which contribute to restoration of those habitats. Keeping desired wetland status by means of water management was possible when the factors influencing catchment runoff process have been recognized. The example of the catchment, in which the swamps and the peatlands have gained high protection status, is the Biebrza River, located in North-Eastern Poland. The factors which influence the runoff of the Biebrza River alone, were subject to a wide range of research, while the runoff process in the Biebrza tributaries has not yet been well recognized. In the Upper Basin of the Biebrza River the main tributaries are: the Sidra River and the Kamienna River. For the catchment areas of those rivers WetSpa model was applied for runoff simulation basing on soil-atmosphere-plant mass balance at a catchment scale. Hydrological processes simulated included: precipitation, evapotranspiration, plant canopy interception, soil interception, infiltration and capillary rise, ground water flow. Global model parameters were calibrated for the catchment of the Kamienna River and for the catchment of the Sidra River and the comparison was made for the values of global parameters for each catchment. The model quality was verified on the independent data set. The model performance was estimated to be satisfactory for high flows, but unsatisfactory for low flows in both catchments. The differences in the runoff process for

L. Porretta-Brandyk (✉) · J. Chormański · T. Okruszko
Department of Hydraulic Engineering and Environmental Restoration,
Warsaw University of Life Sciences, Nowoursynowska Street 159,
02-776 Warsaw, Poland
e-mail: Laurap@levis.sggw.pl

A. Brandyk
Department of Environmental Improvement, Warsaw University of Life Sciences,
Nowoursynowska Street 159, 02-776 Warsaw, Poland

D. Mirosław-Świątek and T. Okruszko (eds.), *Modelling of Hydrological Processes in the Narew Catchment*, Geoplanet: Earth and Planetary Sciences,
DOI: 10.1007/978-3-642-19059-9_3, © Springer-Verlag Berlin Heidelberg 2011

both catchments concerned climatic factors, influencing evapotranspiration, precipitation and snow thawing.

1 Introduction

The valley of the Biebrza River, located in north-eastern Poland, is considered as a unique place on the European scale due to exceptional natural values of wetlands and peatlands located within that valley (Okruszko 1990). Preserving the desired environmental status of the Biebrza wetlands is important for the quality and the quantity of surface water resources of Biebrza as well as its' tributaries. The runoff process of the Biebrza River has currently been the subject of a wide research (Wassen et al. 2006; Okruszko et al. 2006; Mirosław-Świątek and Chormański 2007; Mirosław-Świątek et al. 2007; Szporak et al. 2008; Chormański et al. 2009; Kardel et al. 2009), with considerable part of research performed for the Upper Biebrza Basin (Wasilewski and Chormański 2009; Van Loon et al. 2009; Chormański and Batelaan, submitted). Still there is an insufficient research on hydrological regime of Biebrza tributaries. In this paper, the research was focused on river runoff modeling for two catchments which are the tributaries to Biebrza in the upper reach: Sidra and Kamienna rivers. The outflow hydrographs for both catchments were simulated using WetSpa computer software, which enables rainfall-runoff modeling on a catchment scale on the basis of mass and energy balance in soil-plant-atmosphere system (Batelaan et al. 1996; Wang et al. 1996). WetSpa is GIS based distributed hydrological model, partly physically based. Since physically based distributed hydrological modelling on catchment scale requires many input layers and generate spatially distributed outputs, GIS is a very useful tool because of its efficiency in data storage, display and maintenance. Distributed hydrological models usually use GIS capabilities to estimate spatial parameters from topography and digital maps of soil type and land-use. The powerful GIS tools give new possibilities for hydrological research in under-standing the fundamental physical processes underlying the hydrological cycle and the solution of mathematical equations representing those processes. Except of WetSpa, many hydrological models with a flood prediction component have been developed or updated to use DEM's, such as the model SWAT (Arnold et al. 1998; Arnold and Fohrer 2005), CASC2D, DWSM (Borah et al. 2002), HYDROTEL (Fortin et al. 2001), whereas models like SHE (Abbott et al. 1986) and TOP-MODEL (Beven 2001) were adapted to benefit from GIS data (Quinn et al. 1991; Ewen et al. 2000). These models are either partly or entirely coupled with GIS.

The automated calibration and validation of the model was executed, and the assessment of that crucial processes was done by calculating errors of mathe-matical models performance. Then, the comparison was made for error values for both modeled catchments and the differences and similarities of global model parameters for both catchments were discussed. Error values have shown that the

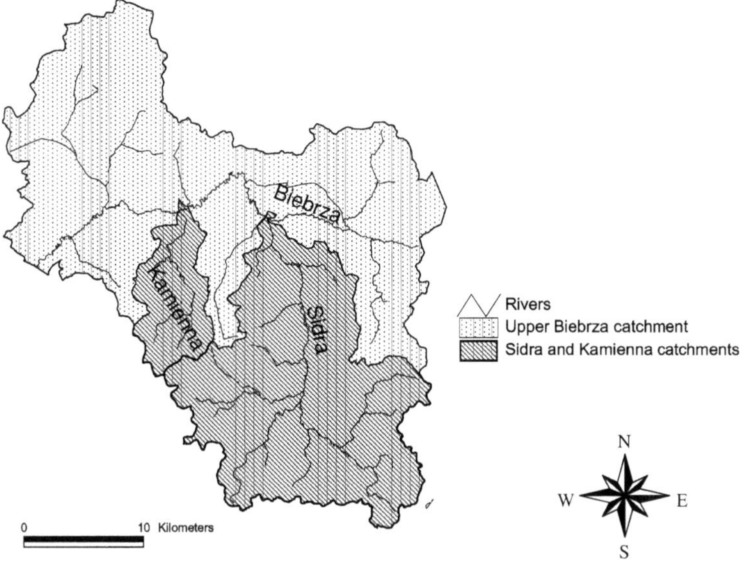

Fig. 1 Catchment of the Upper Biebrza River and the catchments of Kamienna and Sidra

model efficiency was estimated to be good for high flows, but unsatisfactory for low flows in both catchments. The differences in the runoff process for both catchments concerned climatic factors, influencing evapotranspiration, precipitation and snow thawing.

2 Research Area

The research catchments were selected in north-eastern Poland: the Sidra River closed by Harasimowicze gauge, and the Kamienna River, with Kamienna Stara gauge (Fig. 1). Both rivers are one of the main tributaries left side to the Biebrza River. Both catchments are located in the Upper Biebrza Basin—a river valley which is a part of the Biebrza National Park—added to the RAMSAR Convention list as one of the most important worldwide wetlands. The valley has been formed as an ice marginal valley and is relatively long (40 km) and narrow (2–3 km). It is filled with the thick deposits of peat (usually 2–5 m) partly underlain by gyttja layer (1–4 m) (Żurek 1994).

The Upper Biebrza river catchment is located in the subcontinantal climate zone and has a yearly average temperature of 6.8°C. An average annual precipitation ranges from 550 to 700 mm year^{-1}. A maximum precipitation has been noted in summer months (July and August) and in the Biebrza valley is equal to 65–70% of the yearly total (this is only 60% in the surrounding moraine plateaus (Kossowska Cezak 1984). The evapotranspiration is between 460 and 480 mm year^{-1} (Kossowska Cezak 1984). In spring and early summer snowmelt occurs, playing an important role

Table 1 Physico-geographical characteristics of Kamienna and Sidra catchments

Catchment	Area (km^2)	River length (km)	Slope (‰)
Kamienna	56	17.9	3.00
Sidra	264	32.8	1.46

Table 2 Land use structure of Kamienna and Sidra catchments

Catchment	Land use (%)			
	Arable grounds	Forests and shrubs	Grasslands	Urbanized and communiaction areas
Kamienna	71.2	13.7	14.1	1.0
Sidra	73.8	24.9	1.0	0.3

in generating spring floods. Snow processes in the Upper Biebrza Basin represent a significant component of the hydrologic cycle and need to be considered in simulation of the hydrologic processes.

Basic physico-geographical characteristics of selected catchments are given in Table 1. The catchment area of the Kamienna River is five times smaller than the catchment area of the Sidra River, the total length of Kamienna is two times smaller, while the slope is approximately two times higher than for Sidra catchment.

Both catchments are used for agriculture in about 70% of their area. However, the catchment of Kamienna, in comparison to Sidra catchment, is characterized by a smaller surface of forests and shrubs with a relatively high share of grasslands with typically high retention capabilities (Table 2). Hydrological regime of the catchment is described in Table 3.

3 Methods

A model selected for this work was the GIS-based distributed watershed model—WetSpa (developed for simulation of rainfall-runoff processes in the catchment scale). WetSpa is a grid-based hydrological model for water and energy transfer between soil, plants and atmosphere, which was originally developed by Wang et al. (1996) and adopted for flood prediction by de Smedt et al. (2000), Liu et al. (2002). The model was successfully tested for flood hydrograph calculation in urbanized catchments (Chormański et al. 2008) and for lowland catchments with significant contribution of organic soils (Chormański and Batelaan 2010, submitted). In an actual version a model was developed as an ArcGIS 9.2 module by Chormański and Michalowski (2010, submitted). For each grid cell a vegetation, root, transmission and saturated zone is considered in the vertical direction (Fig. 2). The hydrological processes parameterized in the model are: precipitation, interception, depression storage, surface runoff, infiltration, evapotranspiration, percolation, interflow and groundwater flow. The total water balance for a raster

Table 3 Characteristic discharges and unit runoff values for both catchments, estimated for the period 1973–1987

Cathcment	HHQ $(m^3\ s^{-1})$	MMQ $(m^3\ s^{-1})$	MEQ $(m^3\ s^{-1})$	LLQ $(m^3\ s^{-1})$	HHq $(l\ s^{-1}\ km^{-2})$	MMq $(l\ s^{-1}\ km^{-2})$	MEq $(l\ s^{-1}\ km^{-2})$	LLq $(l\ s^{-1}\ km^{-2})$
Kamienna	28.5	0.5	0.2	0.0	508.9	8.9	3.0	0.0
Sidra	26.0	1.5	1.1	0.2	98.5	5.6	4.2	0.7

Explanation for Table 3: HHQ, the highest of the high observed discharges; MMQ, the mean of the mean observed discharges; MEQ, the median of the mean observed discharges; LLQ, the lowest of the low observed discharges

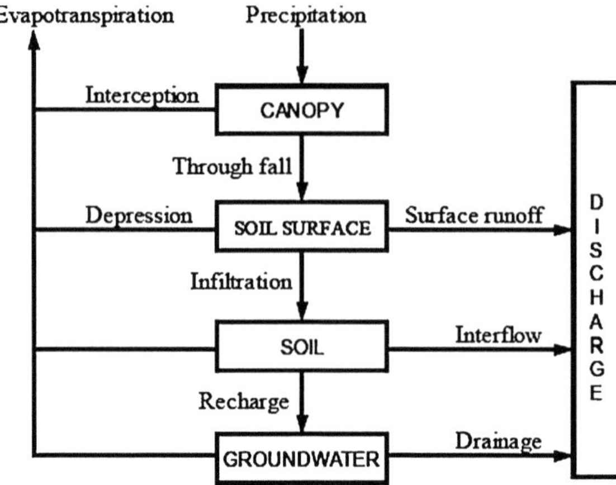

Fig. 2 The structure of WetSpa at grid cell level

cell is composed of the water balance for the vegetated, bare soil, water and impervious parts of each cell. That makes possible to represent within-cell heterogeneity of the land cover. A mixture of physical and empirical relationships is used to describe the hydrological processes in the model. Interception reduces the precipitation to net precipitation, which on the ground is separated into rainfall excess and infiltration. Rainfall excess is calculated using a moisture-related modified rational method with a potential runoff coefficient depending on land cover, soil type, slope, rainfall intensity, and antecedent moisture content of the soil. The calculated rainfall excess fills the depression storage at the initial stage and runs off the land surface simultaneously as overland flow (Linsley et al. 1982).

The infiltrated part of the rainfall stays as soil moisture in the root zone, moves laterally as interflow or percolates as groundwater recharge depending on the moisture content of the soil. Both percolation and interflow are assumed to be gravity driven in the model. Percolation out of the root zone is determined by the hydraulic conductivity, which is dependent on the moisture content as a function of the soil pore size distribution index:

$$R = K(\theta) = K_s \left[\frac{\theta - \theta_r}{\theta_r - \theta_s} \right]^{\left(\frac{2+3B}{B} \right)} \tag{1}$$

where $K(\theta)$, unsaturated hydraulic conductivity (mm h^{-1}); θ, soil moisture content (mm h^{-1}); θ_r, residual moisture content (m^3 m^{-3}); θ_s, soil porosity (m^3 m^{-3}) B, cell pore size distribution index (–).

Interflow is assumed to occur in the root zone after percolation and becomes significant only if the soil moisture is higher than field capacity. Darcy's law and a kinematic approximation are used to estimate the amount of interflow generated from each cell, in function of hydraulic conductivity, moisture content, slope, and root depth. Interflow is assumed to contribute to the nearest channel or ditch flows (Liu et al. 2002):

$$F = \frac{C_f \cdot D \cdot S_o K(\theta)}{W} \tag{2}$$

where D(L) is the root depth; S_o, the surface slope (L L^{-1}); W, cell width (L); C_f(–) is scaling parameter (a function of landuse) to consider river density and effects of organic matter on the horizontal hydraulic conductivity in the top soil layer.

In this case study, as only one measuring station was available, this method was not used. Due to limited knowledge about the bedrock, a non-linear relationship is used to estimate groundwater flow on each subcatchment. The total discharge at the outlet is the sum of the overland flow, interflow and groundwater flow from all the grid cells.

Potential evapotranspiration for the vegetation period (1 April–30 September) was calculated according to the Penman type formula, modified for Polish condition by Roguski et al. (1988) based on meteorological data recorded at the Biebrza Meteorological Station, located in the Middle Biebrza Basin. This formula can be written in the following form:

$$ET_0 = \left[G_0(1-a)\left(0.209 + 0.565\frac{S}{S_0}\right) - \sigma T^4\left(0.56 - 0.08\sqrt{e}\right)\left(0.10 + 0.90\frac{S}{S_0}\right) \right]$$
$$\frac{1}{59}\frac{F_t}{F_t + 0.65} + (e_w - e)(1 + 0.4v)\frac{0.26}{1 + \frac{F_t}{0.65}} \tag{3}$$

where ET_0, reference crop evapotranspiration (mm d^{-1}); G_0, the amount of radiation received at the top of the atmosphere (cal cm^{-2} d^{-1}); a, albedo (–); S, actual sunshine (h); S_0, maximum possible sunshine (h); T, mean daily air temperature (°K); σ, Stefan–Boltzman constant equal 1.18×10^{-7} (cal cm^{-2} d^{-1} °K); e, mean actual vapour pressure (hPa); e_w, saturation vapor pressure at mean air temperature (hPa); F_t, the slope of the saturation water pressure–temperature curve at air temperature (hPa); v, mean wind speed at 10 m (m s^{-1}).

The actual evapotranspiration from soil and plants is calculated for each grid cell using the relationship developed by Thornthwaite in 1959 and as a function of potential evapotranspiration, vegetation stage, and moisture content in the cell. Outside of the vegetation period potential evapotranspiration is assumed to be zero.

Runoff from different cells in the watershed is routed to the watershed outlet depending on flow velocity and wave coefficient by using the diffusive wave approximation method. An approximate solution proposed by De Smedt et al. (2000) in the form of an instantaneous unit hydrograph (IUH) was used in the model, relating the discharge at the end of a flow path to the available runoff at the start of the flow path:

$$u(t) = \frac{1}{\sigma\sqrt{2\pi t^3/t_0^3}}\exp\left[\frac{(t_o - t)^2}{4\sigma^2 t/t_o}\right] \tag{4}$$

where $u(t)$, the unit response function (T^{-1}); t_0, the average travel time (T) to the outlet along the flow path and σ is the standard deviation of the flow time. Parameters t_0 and σ are spatially distributed, and can be obtained by integration

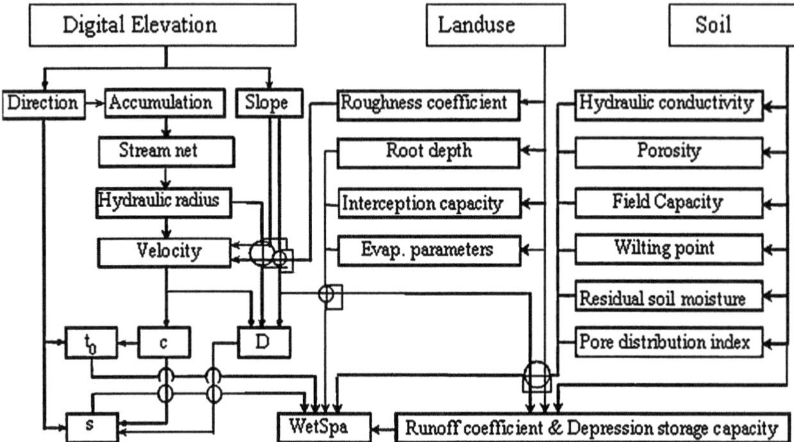

Fig. 3 Parametrization of local parameters in WetSpa model (Liu 2004)

along the topographically determined flow paths as a function of flow celerity and dispersion. These parameters are physically based and can be estimated by using standard GIS functions.

The total flow hydrograph at a basin outlet can be obtained by a convolution integral of the flow response from all grid cells.

$$Q(t) = \int\limits_A \int\limits_o^t Q_o(\tau)U(t-\tau)d\tau dA \tag{5}$$

where $Q(t)$, the outlet flow hydrograph ($L^3\,T^{-1}$); Q_O, the excess precipitation in a grid cell ($L\,T^{-1}$); τ (T), the time delay and A (L^2) a drainage area of the watershed.

Although the spatial variability of land use, soil and topographic properties within a watershed are considered in the model, the groundwater response is modelled on sub-catchment scale. The simple concept of a linear reservoir is used to estimate groundwater discharge, while a non-linear reservoir method is optional in the model (Wittenberg and Sivapalan 1999). The groundwater outflow is added to the generated runoff to produce the total stream flow at the sub-watershed outlet. Time-dependent inputs of the model are precipitation and potential evapotranspiration. Model parameters such as interception and depression storage capacity, potential runoff coefficient, overland roughness coefficient, root depth, soil property parameters, average travel time to the outlet, dispersion coefficient, etc., are determined for each grid cell using lookup tables and a high-resolution DEM, soil type and land-use maps (Fig. 3). The main outputs of the model are river flow hydrographs, which can be defined for any location along the channel network, and spatially distributed hydrological characteristics, such as soil moisture, infiltration rates, groundwater recharge, surface water retention, runoff, etc.

During the calibration 11 global parameters are used by the model. Those parameters are listed below:

Ki—interflow scaling parameter,
Kg—the groundwater recession coefficient
K_ss—relative soil moisture
K_ep—plant coefficient for estimating actual potential evapotranspiration
GO—Initial groundwater storage (m)
G_max—Maximum groundwater storage (m)
TO—Base temperature for estimating snow melt (°C)
K_snow—degree day coefficient for calculating snow melt (mm/mm/°C/day)
K_rain—rainfall degree day coefficient (mm/mm/°C/day),
K_run—an exponent reflecting the effect of rainfall intensity on the actual runoff coefficient when the rainfall intensity is very small
P_max—threshold rainfall intensity (mm/day)

3.1 Model Calibration and Evaluation

The rough calibration is performed automatically with PEST, a model-independent non-linear parameter estimator, which is implemented in WetSpa (Liu et al. 2005). Then, a fine-tuning calibration is performed manually. The calibration processes consist of running the model as many times as it needs to adjust model parameters within their predetermined range until the discrepancies between model outputs and a complementary set of flow observations is reduced. The model is evaluated by graphical comparison of simulated and observed hydrographs and assessment of goodness of fit between them by five statistical evaluation criteria CR1, CR2, CR3, CR4, and CR5:

3.1.1 CR1: Model Bias

Model bias can be expressed as the relative mean difference between predicted and observed stream flows for a sufficiently large simulation sample, reflecting the ability of reproducing water balance, and perhaps the most important criterion for comparing whether a model is working well in practice. Error CR1 is the model bias, Q_{si} and Q_{oi} are the simulated and observed stream flows at time step i (m^3/s), and N is the number of time steps over the simulation period. Model bias measures the systematic under or over prediction for a set of predictions. A lower CR1 value indicates a better fit, and the value 0.0 represents the perfect simulation of observed flow volume. The criterion is given by the equation:

$$CR1 = \frac{\sum_{i=1}^{N}(Q_{si} - Q_{oi})}{\sum_{i=1}^{N}Q_{oi}} \tag{6}$$

3.1.2 CR2: Model Confidence

Model confidence is important criterion in assessment of continuous model simulation, and can be expressed by determination coefficient. CR2 represents the proportion of the variance in the observed discharges that are explained by the simulated discharges. It varies between 0 and 1, with a value close to 1 indicating a high level of model confidence. It is given by the equation:

$$\mathrm{CR2} = \frac{\sum_{i=1}^{N} \left(Q_{si} - \overline{Q_{oi}} \right)^2}{\sum_{i=1}^{N} \left(Q_{oi} - \overline{Q_o} \right)^2} \tag{7}$$

3.1.3 CR3: Nash–Sutcliffe Efficiency

The Nash–Sutcliffe coefficient (Nash and Sutcliffe 1970) describes how well the stream flows are simulated by the model. This efficiency criterion is commonly used for model evaluation, because it involves standardization of the residual variance, and its expected value does not change with the length of the record or the scale of runoff. The Nash–Sutcliffe efficiency is used for evaluating the ability of reproducing the time evolution of stream flows. The CR3 value can range from a negative value to 1, with 1 indicating a perfect fit between the simulated and observed hydrographs. CR3 below zero indicates that average measured stream flow would have been as good a predictor as the modelled stream flow. A perfect model prediction has CR3 score equal to 1. It has the form of:

$$\mathrm{CR3} = \frac{\sum_{i=1}^{n} \left(Qs_i - Qo_i \right)^2}{\sum_{i=1}^{n} \left(Qo_i - Qo \right)^2} \tag{8}$$

3.1.4 CR4 Logarithmic Version of Nash–Sutcliffe Efficiency for Low Flow Evaluation

CR4 is a logarithmic Nash–Sutcliffe efficiency for evaluating the ability of reproducing the time evolution of low flows, and e is an arbitrary chosen small value introduced to avoid problems with nil observed or simulated discharges. A perfect value of CR4 is 1.

$$\mathrm{CR4} = \frac{\sum_{i=1}^{n} \left[\ln(Qs_i + \varepsilon) - \ln(Qo_i + \varepsilon) \right]^2}{\sum_{i=1}^{n} \left[\ln(Qo_i + \varepsilon) - \ln(Qo + \varepsilon) \right]^2} \tag{9}$$

Table 4 Model performance categories to indicate the goodness level

Range of R^2	Range of C1	Range of C2–C5	Model quality category
$0.99 \leq R^2 < 1.00$	<0.05	>0.85	Excellent
$0.95 \leq R^2 < 0.99$	0.05–0.10	0.65–0.85	Very good
$0.90 \leq R^2 < 0.95$	0.10–0.20	0.50–0.65	Good
$0.85 \leq R^2 < 0.90$	–	–	Fairly good
$0.80 \leq R^2 < 0.85$	–	–	Average
$0.70 \leq R^2 < 0.80$	0.20–0.40	0.20–0.50	Satisfactory/poor
$R^2 < 0.70$	>40	<20	Unsatisfactory/very poor

3.1.5 CR5 Adapted Version of Nash–Sutcliffe Efficiency for High Flow Evaluation

CR5 is an adapted version of Nash–Sutcliffe criterion for evaluating the ability of reproducing the time evolution of high flows. A perfect value of CR5 is 1. To evaluate the goodness of the model performance during calibration and validation periods, the intervals listed in Table 4 have been adopted (Andersen et al. 2001). These criteria are not of the fail/pass type, but evaluate the performance in the categories from excellent to very poor.

$$\text{CR5} = \frac{\sum_{i=1}^{n}(Qo_i + Qo)(Qs_i - Qo_i)^2}{\sum_{i=1}^{n}(Qo_i + Qo)(Qo_i - Qo)^2} \tag{10}$$

Additionally, the determination coefficient was calculated for the estimation of model quality, according to the criteria given in Table 4 (Ostrowski 1994):

$$R^2 = \frac{\sum_{1}^{n}(\hat{y}_t - \bar{y})^2}{\sum_{1}^{n}(y_t - \bar{y})^2} \tag{11}$$

where y_t, true value of a variable at time t; $\hat{y}_t$, modeled value of a variable at time t; $\bar{y}_t$, arithmetical mean of a variable.

After model calibration and verification, the comparison of global model parameter values for Sidra and Kamienna catchments was performed in order to discover differences and similarities in the formation of runoff process in lowland catchments. Model efficiency for low flows was estimated, which, in small catchments usually reached unsatisfactory values according to previous inquiries (Liu et al. 2005). The sensitivity calculation of calibrated model parameters was also performed.

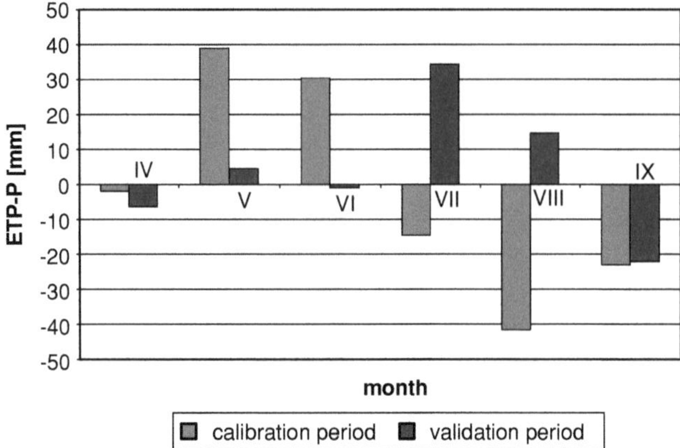

Fig. 4 Monthly difference between potential evapotranspiration and precipitation calculated as monthly average in calibration period 1.XI.1978–31.X.1980 and validation period 01.XI.1981–31.X.1983, respectively

3.2 Input Data

The input data set is historical meteorological and hydrological data for the period 1976–1986. The input data included daily precipitation and mean daily temperature from Rozanystok station, located in the Upper Biebrza Basin in the distance of 15 km from the computational river cross-sections, and potential evapotranspiration from the Biebrza Meteorological Station, located in the Middle Biebrza Basin, in the distance of 30 km from computational cross-sections: Harasimowicze and Kamienna Stara. Meteorological services on both stations in Rogożynek and in Biebrza is provided by Institute of Meteorology and Water Management (IMGW).

The correctness of performed modeling was evaluated on the basis of daily discharges for the analysed gauging cross-sections. The range of data was divided into two 5-year periods: the data for 1977–1981 was used for model calibration while the data for 1982–1986 provided for model validation. Simulations were performed for a daily time step. Figure 4 shows water deficit in both periods as a difference between potential evapotranspiration and rainfall calculated as monthly average in calibration period 1.XI.1978–31.X.1980 and validation period 01.XI.1981–31.X.1983, respectively. In both periods we observed that evapotranspiration is much higher than precipitation in May and June of calibration period and in July and August of validation period. The amount of water losses can strongly influence water balance and resulted as smaller runoff reaction due to dry condition in catchments.

Table 5 Global model parameters for Sidra catchment

Parameter	Value			% change
	Range	Initial value	Final value (calibrated)	
Ki	0–10	1.5	3.466	+43.3
Kg	0–0.05	0.005	0.00348	−30.4
Kss	0–2	1.5	1.373	−9.3
Kep	0–2	1.0	0.873	−14.5
G0	0–100	0.2	0.2	0
Gmax	0–2,000	200	197.447	−1.3
TO	0–1	0.7	0.466	−33.4
K_snow	0–10	1.5	3.5	−8.6
K_rain	0–0.05	0.0005	0.0005	0
K_run	0–5	2	2	0
P_max	0–500	500	70	−714.3

Table 6 Global model parameters for Kamienna catchment

Parameter	Value			% change
	Range	Initial value	Final value (calibrated)	
Ki	0–10	3.0	1.956	−34.8
Kg	0–0.05	0.013	0.004	−69.2
Kss	0–2	2.0	1.457	−27.2
Kep	0–2	2.0	1.439	−28.1
G0	0–100	0.1	0.008	−92.0
Gmax	0–2,000	200	201.714	+0.84
TO	0–1	0.7	0.694	−0.85
K_snow	0–10	1.5	1.781	+15.7
K_rain	0–0.05	0.0005	0.0003	−40.0
K_run	0–5	2	2.231	+11.6
P_max	0–500	500	120	−416

4 Results

Global model parameters for Kamienna and Sidra catchments are given in Tables 5 and 6.

The final, calibrated values of model parameters for Sidra catchment were smaller in comparison to the initial values, except for: G0 (Initial groundwater storage), K_rain (rainfall degree day coefficient) and K_run (an exponent reflecting the effect of rainfall intensity on the actual runoff coefficient when the rainfall intensity is very small), which did not change during calibration.

The value of Ki (interflow scaling parameter) was increased during calibration process. The calibrated value of that parameter indicates that lateral flow through

Table 7 Calibrated values of model parameters for Sidra and Kamienna catchments

Parameter	Values	
	Sidra catchment	Kamienna catchment
Ki	3.466	1.956
Kg	0.00348	0.004
Kss	1.373	1.457
Kep	0.873	1.439
G0	0.2	0.008
Gmax	197.447	201.714
TO	0.466	0.694
K_snow	3.5	1.781
K_rain	0.0005	0.0003
K_run	2	2.231
P_max	70	120

soil will have a higher share in runoff formation than the vertical soil flow. The P_max value (threshold rainfall intensity) was subject to the highest change. The decrease of the value of Kg coefficient (groundwater recession coefficient) implies a lower ground water retention in comparison to the initial one. The values of Kep (plant coefficient for estimating actual evapotranspiration) and T0 (base temperature for estimating snow melt) were also decreased.

The values of the parameters for Kamienna catchment were decreased during calibration, except for G max (Maximum groundwater storage), K_snow (degree day coefficient for calculating snow melt) and K_run (an exponent reflecting the effect of rainfall intensity on the actual runoff coefficient when the rainfall intensity is very small), which were increased in comparison to their initial values. The value of P_max (threshold rainfall intensity) had the highest change. The calibrated value of Ki (interflow scaling parameter) proves that the lateral flow of water through soil will have a higher input to the formation of the runoff than the vertical soil flow.

Table 7 illustrates calibrated parameter values for both catchments, used for performing outflow simulations.

The values of Ki parameter indicate that for both catchments the higher share in runoff formation has the lateral soil flow than the vertical one. The higher anisotropy of ground water runoff parameters is characteristic for Sidra catchment. It is mostly caused by the differences in soil composition and structure. The ground water recession factor (Kg) recorded similar values for both catchments, as well as the values of relative soil moisture (Ks). Higher potential evapotranspiration values (adjusted by Kep parameter) will occur in Kamienna catchment, despite the similarities of land use and vegetation cover of the analysed catchments. Higher values of initial ground water storage (G0) were obtained for Sidra catchment, which is mostly caused by hydro-meteorological conditions prevailing in the chosen research period and also by the differences in soil structure. Maximum ground water storage (Gmax) has similar values for both catchments, contrary to:

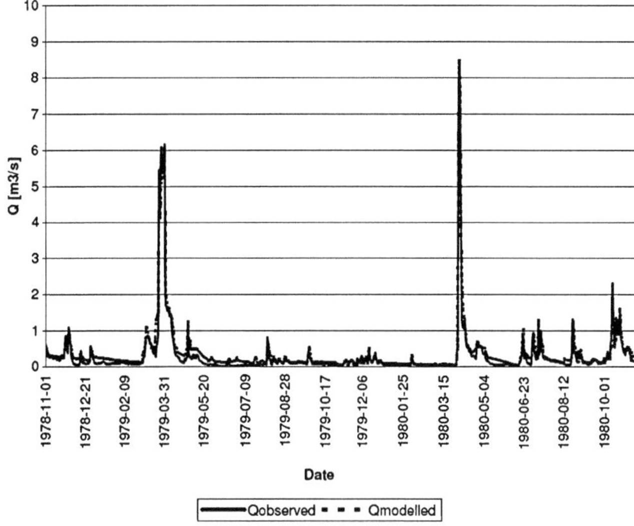

Fig. 5 Calculated and observed daily discharge hydrograph for the Kamienna River, calibration period 1.XI.1978–31.X.1980

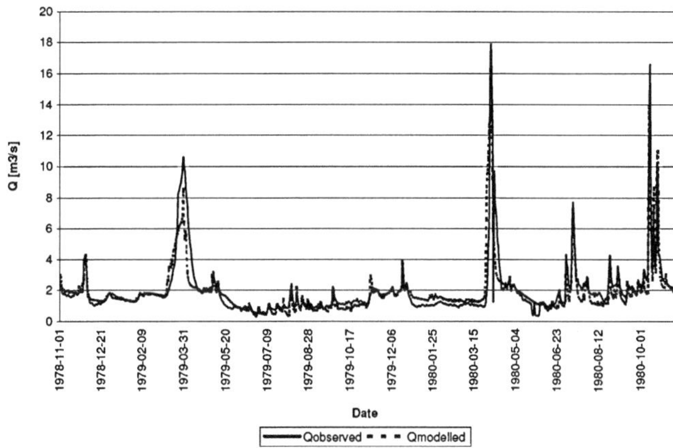

Fig. 6 Calculated and observed daily discharge hydrograph for the Sidra River, calibration period 1.XI.1978–31.X.1980

TO—base temperature for estimating snow melt and K_snow—degree day coefficient for calculating snow melt. Those parameter values confirm an earlier beginning of snow thawing in the Sidra catchment and higher susceptibility of snow cover to thawing in that catchment. P_max—threshold rainfall intensity reached higher values for Kamienna catchment. It could cause a higher HHq value for Kamienna Stara gauge that for Harasimowicze gauge at the Sidra River apart

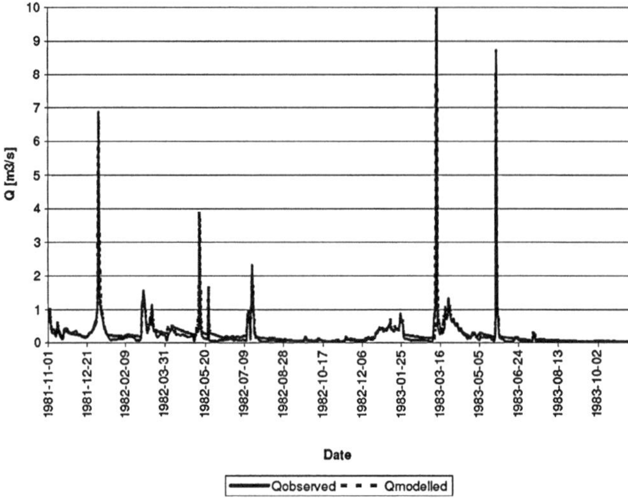

Fig. 7 Calculated and observed daily discharge hydrograph for the Kamienna River, validation period 01.XI.1981–31.X.1983

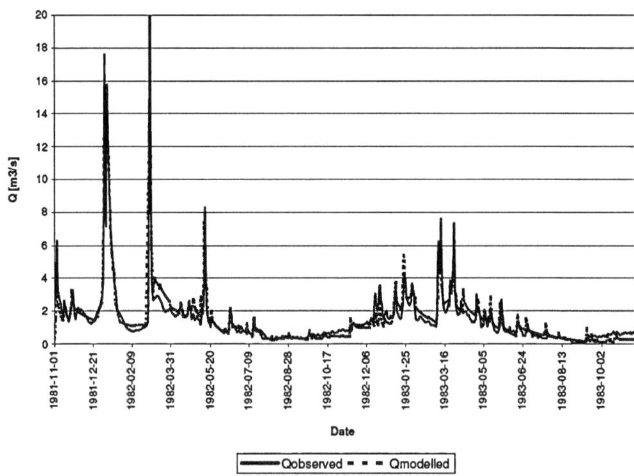

Fig. 8 Calculated and observed daily discharge hydrograph for the Sidra River, validation period 01.XI.1981–31.X.1983

from a higher slope of Kamienna catchment. Figures 5 and 6 show daily discharge hydrographs for two exemplary years chosen from calibration period. The determination coefficient values R^2 were calculated for that period to estimate the fit between the observed and the calculated discharges. Figures 7 and 8 represent daily discharges for two chosen years of validation period. For the Kamienna

Table 8 Calculated parameter sensitivities

Parameter	Kamienna	Sidra
Ki	1.71	1.58E-02
Kg	0.85	1.09
K_ss	3.82	0.69
K_ep	4.17	4.54
G0	4.6 E-04	9.12 E-04
G_max	3.7 E-05	3.25E-05
T0	0	0
K_snow	0	0
K_rain	0	0
K-run	6.21E-03	9.82E-04
P_max	0.45	0.25

Table 9 Nash–Sutcliffe Efficiency C1–C5 for the calibration and verification of WetSpa model for Kamienna Stara gauge, the Kamienna River

Błąd	C1	C2	C3	C4	C5	R^2
Calibration	0.08	0.70	0.72	0.31	0.83	0.80
Verification	0.05	0.73	0.69	0.35	0.81	0.80
Optimum	0	1	1	1	1	1

Table 10 Nash–Sutcliffe efficiency C1–C5 for the calibration and verification of WetSpa model for Harasimowicze gauge, the Sidra River

Parameter	C1	C2	C3	C4	C5	R^2
Calibration	0.06	0.71	0.69	0.20	0.78	0.72
Verification	0.03	0.69	0.66	0.34	0.81	0.69
Optimum	0	1	1	1	1	1

River at Kamienna Stara gauge the value of R^2 is 0.8 while for the Sidra River at Harasimowicze gauge the value of R^2 is 0.72 (Table 8). According to the applied criteria by Ostrowski (1994) the quality of the model for Kamienna catchment was the average and for Sidra catchment it was satisfactory. The sensitivities of model parameters were most significant for K_ss, K_ep and Ki (Table 8).

The comparison of the calculated and the observed discharge hydrographs confirms the proper calibration process. The goodness of fit was also estimated by the errors C1–C5 (Nash–Sutcliffe Efficiency) (Tables 9 and 10).

The values of C1–C5 errors were similar for both catchments. The values of C1 and C2, C3, C5 were estimated to be close to optimum (Liu et al. 2005)—and show very good model performance. The values of C4 were low for both catchments and were not coherent with the optimum. Similar C1–C5 values were obtained for the WetSpa model of the whole Upper Basin up to the closing cross-section at Sztabin, constructed in a lower resolution and including both analysed subcatchments (Chormański and Batelaan 2010). For the WetSpa model of the whole Upper Basin an appropriate high values of C1, C2, C3 and C5 were reached while the value of C4 was poor as for Sidra and Kamienna catchments.

5 Conclusions

Global model parameter values were typical for lowland catchments.

The differences in runoff forming process result from climatic factors, even if both catchments are located within one similar climatic region of Poland.

The differences concern climatic processes such as: evaporation, snow cover thawing and threshold rainfall intensities evoking floods.

Ground water runoff parameters are similar for both catchments, the only difference is the higher anisotropy of soils conductivity in the catchment of Sidra.

The determination coefficient proved a satisfactory quality of the model for both catchments.

Model quality is good for daily discharges and in general for high flows. It is not satisfactory for low flows, which was also reported in previous investigations for small lowland catchments.

The desired quality of calibration of rainfall-runoff model parameters for small lowland catchments is reached in an automated manner followed by manual refinement of parameter values.

Calibration enabled to determine such values of parameters describing runoff process, that lead to the satisfactory reproduction of time evolution of flows in the studied lowland catchments.

Moving from large-scale models to local ones did not successfully solve the problem of quality of low flows simulation.

Acknowledgments Financial support was received from the Polish Ministry of Science and Higher Education as part of the research project N30505232/1917 is greatly appreciated.

References

Abbott MB, Bathurst JC, Cunge JA, O'Connell PE, Rasmussen J (1986) An introduction to the European hydrological system, SHE, 1: history and philosophy of a physically base, distributed modelling system. J Hydrol 87:45–59

Andersen J, Refsgaard J, Jensen KH (2001) Distributed hydrological modelling of the senegal river basin—model construction and validation. J Hydrol 247:200–214

Arnold JG, i Fohrer N (2005) SWAT2000: current capabilities and research opportunities in applied watershed modeling. Hydrol Process 19(3):563–572

Arnold JG, Srinivasan R, Muttiah RS, Wiliams JR (1998) Large area hydrologic modeling and assessment: Part I. Model development. J Am Water Resour Assoc 34(1):73–89

Batelaan O, Wang ZM, De Smedt F (1996) An adaptive GIS toolbox for hydrological modeling. In: Kovar K, Nachtnebel HP (eds.) Application of geographic information systems in hydrology and water resources management, IAHS Publ. No. 235

Beven K (2001) Rainfall-runoff modeling: the primer. Wiley, Hoboken pp 360

Borah DK, Xia R, Bera M (2002) DWSM—a dynamic watershed simulation model. In: Singh VP, Freyert DK (eds) Mathematical models of small watershed hydrology and applications. Water Resources Publications, LLC, Highlands Ranch, pp 113–166

Chormański J, Batelaan O (2010) Application of the WetSpa distributed hydrological model for catchment with signiphicant contribution of organic soil. Upper Biebrza case study. Annals of Warsaw University of Life Sciences—SGGW Land Reclamation (submitted)

Chormański J, Michałowski R (2010) ArcGIS extension for Wetspa simulator Przegląd Naukowy Inżynieria i Kształtowanie Środowiska, in Polish (submitted)

Chormański J, Van de Voorde T, De Roeck T, Batelaan O, Canters F (2008) Improving distributed runoff prediction in urbanized catchments with remote sensing based estimates of impervious surface cover. Sensors 8:910–932

Chormański J, Mirosław-Świątek D, Michałowski R (2009) A hydrodynamic model coupled with GIS for flood characteristics analysis in the Biebrza riparian wetland. Oceanol Hydrobiol Stud 38(1):65–73

De Smedt F, Liu YB, Gebremeskel S (2000) Hydrologic modeling on a catchment scale using GIS and remote sensed land use information. In: Brebbia CA (ed) Risk analysis, 2nd edn. WTI press, Southampton, pp 295–304

Ewen J, Parkin G, O'Connell PE (2000) SHERTRAN: distributed river basin flow and transport modelling system. J Hydrol Eng 5:250–258

Fortin JP, Turcotte R, Massicotte S, Moussa R, Fitzback J, Villeneuve JP (2001) A distributed watershed model compatible with remote sensing and GIS data, I: description of the model. J Hydrol Eng ASCE 6(2):91–99

Kardel I, Mirosław-Świątek D, Chormański J, Okruszko T, Wassen M (2009) Water management decision support system for Biebrza National Park. Environ Prot Eng 35(2):173–180

Kossowska Cezak U (1984) Climate of the Biebrza ice-margin Halley. Polish Ecol Stud 10(3–4):253–270

Linsley RKJ, Kohler MA, Paulhus JLH (1982) Hydrology for engineers, 3rd edn. McGraw-Hill, New York, p 237

Liu Y (2004) A GIS-based hydrologic model for flood prediction and watershed management. Documentation and user manual. Department of Hydrology and Hydarulic Engineering, Vrije Universiteit, Brussel, p 315

Liu YB, Gebremeskel S, De Smedt F, Pfister L (2002) Flood prediction with the WetSpa model on a catchment scale. In: Wu BS, Wang ZY, Wang GQ, Huang GH, Fang HW, Huang JC (eds) Flood defence 2002. Science Press, New York

Liu YB, Batelaan O, De Smedt F, Poorova J, Velcicka L (2005) Automated calibration applied to a GIS-based flood simulation model using PEST. In: van Alphen J, van Beek E, Taal M (eds) Floods, from defence to management. Taylor-Francis Group, London, pp 317–326

Mirosław-Świątek D, Chormański J (2007) The verification of the numerical river flow model by use of remote sensing. In: Okruszko T, Maltby E, Szatyłowicz J, Mirosław-Świątek D, Kotowski W (eds) Wetlands: Monitoring, Modelling, Management. Taylor &Francis/ Balkema, The Netherlands, pp 173–180

Mirosław-Świątek D, Szporak S, Chormański J, Ignar S (2007) Influence of a different land use on a flood extent in the lower Biebrza valley. The fifth international symposium on environmental hydraulics (ISEH V) Tempe, Arizona, the Grand Canyon State, 4–7 December 2007, s.1–7

Nash J, Sutcliffe J (1970) River flow forecasting through conceptual models, part 1—a discussion of principles. J Hydrol 10:282–290

Okruszko H (1990) Wetlands of the Biebrza Valley their value and future management. Warszawa, 107 p

Okruszko T, Chormański J, Mirosław-Świątek D (2006) Flooding of the riparian wetlands-interaction between surface and ground water. Biebrza wetlands case study, IAHS-AISH, pp 573–578

Ostrowski J (1994) A regional model of the small catchment "MOREMAZ-1", Materiały Badawcze IMGW, ser. Hydrol.i Oceanom.-17, Warsaw (In Polish)

Quinn P, Beven K, Chevallier P, Planchon O (1991) The prediction of hillslope flow paths for distributed hydrological modeling using digital terrain models. Hydrol Process 5:59–79

Roguski W, Sarnacka S, Drupka S (1988) Guidelines for predicting crop and pasture water needs. Mat Instruktarzowe 66, Falenty IMUZ (In Polish)

Szporak S, Mirosław-Świątek D, Chormański J (2008) Hydrodynamic model of the Lower Biebrza River flow—a tool for assessing the hydrologic vulnerability of a floodplain to management practices. Ecohydrol Hydrobiol 8(2–4):331–337

Van Loon AH, Schot PP, Griffioen J, Bierkens MFP, Batelaan O, Wassen MJ (2009) Throughflow as a determining factor for habitat contiguity in a near-natural fen. J Hydrol 379(2009):30–40

Wang ZM, Batelaan O, De Smedt F (1996) A distributed model for water and energy transfer between soil, plants and atmosphere (WetSpa). Phys Chem Earth 21(3):189–193

Wasilewski M, Chormański J (2009) The shuttle radar topography mission digital elevation model as an alternative data source for deriving hydrological characteristics in lowland catchment. Annals of Warsaw University of Life Sciences—SGGW Land Reclamation, No 41, pp 71–82

Wassen M, Okruszko T, Kardel I, Chormański J, Świątek D, Mioduszewski W, Bleuten W, Querner E, Kahloun El M, Batellan O, Meire P (2006) Eco-hydrological functioning of Biebrza wetlands: lessons for the conservation and restoration of deteriorated wetlands. Ecol Stud, vol 191. Springer-Verlag, Berlin, pp 285–310

Wittenberg H, Sivapalan M (1999) Watershed groundwater balance estimation using streamflow recession analysis and baseflow separation. J Hydrol 219:20–33

Żurek S (1994) Geomorphology of the Biebrza valley. In: Okruszko H, Wassen MJ (eds) Towards protection and sustainable use of the Biebrza Wetlands: exchange and integration of research results for the benefit of Polish-Dutch Joint Research Plan. Report 3A: the environment of the Biebrza Wetlands. Utrecht, pp 15–48

Modeling Hydrological Flow Paths During Snowmelt Induced High Flow Event in a Small Agricultural Catchment

Piotr Banaszuk, Małgorzata Krasowska and Andrzej Kamocki

Abstract End-Member Mixing Analysis (EMMA) was used to identify flow paths and source areas controlling river chemistry during a snow melt induced spring high flow event in an agricultural catchment (187 ha) in NE Poland. EMMA using Ca^{2+}, Mg^{2+}, Cl^-, and H^+ showed that stream chemistry could be explained as a three-component mixture of overland flow, shallow groundwater and soil solution from arable soils along the stream margin and deeper groundwater high in base cations and Si. The temporal variability in the flow pathways and solute sources during snowmelt were controlled by soil frost. From the very early beginning of snowmelt overland and rill flows were the main mechanism of runoff generation because of the low permeability of the frozen ground. Solutes transported along with overland flow had the most pronounced impact on river chemistry during peak discharges, when this runoff component contributed up to 70% of stream discharge. High surface runoff contributions produced a pronounced rise and maximum in streamflow PO_4^{3-}. We found that during investigated snowmelt event only a small percentage of the landscape might be a source of subsurface flow. They were primarily saturated areas well-connected to the drainage network: geomorphic hollows and sideslope benches along the stream margins. Shallow groundwater and soil water discharged from those locations were responsible for pronounced ($>60\%$) export of nitrates. The high concentration of solutes (primarily NO_3^-) in the river outflow suggests that during snowmelt, either fluxes of agricultural contaminants bypassed potential buffers, which could constrain their impact on freshwater ecosystems, or that existing buffers were ineffective in removing the contaminants that moved along shallow hydrological pathways. Thus, the short period of snowmelt flood may be perceived as critical from the river water quality perspective.

P. Banaszuk (✉) · M. Krasowska · A. Kamocki
Technical University of Bialystok, 15-351 Bialystok, ul. Wiejska 45a, Białystok, Poland
e-mail: p.banaszuk@pb.edu.pl

D. Mirosław-Świątek and T. Okruszko (eds.), *Modelling of Hydrological Processes in the Narew Catchment*, Geoplanet: Earth and Planetary Sciences, DOI: 10.1007/978-3-642-19059-9_4, © Springer-Verlag Berlin Heidelberg 2011

1 Introduction

Understanding hydrological pathways is crucial to recognizing and managing the effect of agriculture-derived pollutants on freshwater ecosystems (Soulsby et al. 2002). Research in this area has grown considerably over the past two decades, resulting in several methodological approaches and conceptual models that are helpful to identifying the flow paths and runoff mechanisms that control solute transport from terrestrial to aquatic systems. Hydrological pathways and source areas of contaminants have been widely studied in small, homogenous headwater catchments (Christophersen and Neal 1990; Kendall et al. 1999; Pionke et al. 2000) as well as in major river systems (Smart et al. 1998). Most European studies have studied controls on stream water chemistry in regions characterized by a maritime climate with low intra-annual air temperature amplitudes and a high and rather homogenous precipitation regime, or have focused on solute export in the summer half-year during hydrological events generated by rainfall. In such hydro-meteorological conditions, migration of soluble species was dominated by leaching to the groundwater with mass flow. Therefore, subsurface flow was found to be the main factor responsible for the export of agricultural contaminants, mainly nitrates (Haag and Kaupenjohann 2001). Until recently, relatively little has been reported on hydrological pathways and diffuse agricultural contaminant sources influencing riverine ecosystems under a temperate climate with moderately strong continentalism where the frozen ground favours shallow flow paths by which solutes may be rapidly transported to the stream (Szpikowski et al. 2008). Indeed, in recent years numerous comprehensive studies on hydrological flow paths and export mechanisms of solutes during snowmelt have appeared (Laudon et al. 2004; Piatek et al. 2005; Inamdar and Mitchell 2006), but the majority of them have been conducted in small and rather homogenous, forested catchments. Moreover, despite the fact that some of the studied catchments were located at high elevation sites or in a temperate climatic zone, where in winter air temperature fell well below freezing, little if any soil freezing was found in some of these sites (e.g., Creed et al. 1996; Sickman et al. 2003).

Soil frost has significant influence on physical and biological ecosystem processes. Therefore, its occurrence and extent is of particular concern. It has been found that forest ecosystems, due to the insulating effect of deep snow cover and forest floor litter, are likely to have "patchy" or "granular" soil frost, characterized by higher infiltration rates and negligible surface runoff (Campbell et al. 2005). However, under a colder boreal climate extensive soil frost may facilitate the occurrence of overland flow, as has been demonstrated by Laudon et al. (2004) for forested catchment in northern Sweden. In contrast, in open areas such as agricultural ecosystems, soil frost usually forms a continuous and highly impermeable layer that facilitates the generation of overland and rill flow, bypassing soil and groundwater flow paths (Hart 1963; Shanley and Chalmers 1999). It has been also suggested that soil freeze–thaw events enhance the mineralization rate of soil organic matter and therefore have important effects on the cycling and export of nutrients such as nitrogen and carbon (Wang and Bettany 1993, 1994). In north-eastern USA, widespread soil freezing in 1989 has been regarded to be a main cause of the regional increase in streamwater nitrate concentration observed in the subsequent year (Mitchell et al. 1996).

Since climate and land use strongly affect the runoff pattern and intensity of solute export, it is likely that some observations and conclusions formulated on the basis of investigations carried out in forested catchment may not be fully adequate to describe controls on solute export from agricultural watersheds. The primary objective of the present research was to better understand the flow paths that affect the fluxes of dissolved compounds from a small agricultural catchment during snowmelt. We focused on spring snowmelt, because this is the dominant hydrological event in many moderate and high latitude catchments and, thus, is regarded as a prominent factor influencing the quality of surface waters (Laudon et al. 2004).

2 Materials and Methods

2.1 Study Area

Field works were conducted in the small catchment (187 ha) of no-named permanent stream, left side tributary of Horodnianka River (Fig. 1). The elevation range on the catchment is between 118 and 142 m above sea level. The catchment is rural and dominated by low intensity arable farming and unimproved grassland. Arable lands cover ca. 75%, while grasslands (located mainly in the watercourse valley)—16% of catchment area. Woodlands constitute only 3.5%, and built up areas and wastelands 5.5% of the area. Approximately 60% of agricultural land in the catchment has been artificially drained, mainly by underground pipes.

The catchment is made of ground moraine forming gently rolling hills. Surficial sediments consist of medium deep sands and loamy sands underlined by glacial till, and in some places of light and medium loams. The depth to the till measured using soil cores varies from 0.5 to 2 m along the side slopes and 0.4–0.7 m in the valley-bottom locations. Above 20% of the area is covered with patches of deep sands and gravels that form isolated kame hills, which are the highest elevated forms in the catchment. The valley of the stream is filled with organic sediments, mainly shallow (up to 50 cm), highly decomposed peats and alluvial muds. The valley bottom is narrow and watercourse bed cut it to the depth of 80–100 cm. On the significant area of valley's length, arable lands approach the stream bed on the distance of 5–10 m.

NE Poland is located in a temperate climatic zone with distinctly marked continental and a lesser boreal influence. Mean air temperature is 6.9°C. Monthly average temperatures range from −4.5°C in the coldest month of the year (January) to 17.3°C in the warmest (July). The mean annual precipitation in the region is 587 mm

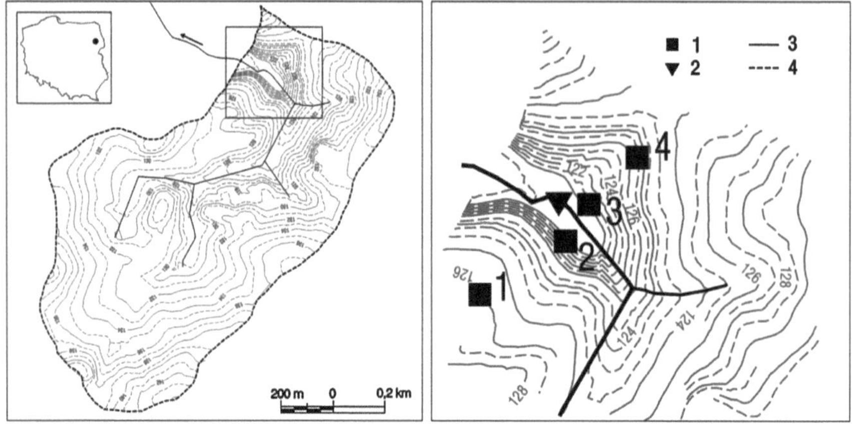

Fig. 1 Study area; *1* monitoring plots, *2* stream gauge and sampling point, *3* hydrographic network, *4* watershed

(1956–2000), of which 60–70% falls between April and September. Permanent snow cover occurs on average on 70–80 days every year, between late December and early March. In winter, the depth of ground frost penetration varies from 30 to 50 cm depending on atmospheric conditions. On average storms occur on 20–25 days a year (Atlas 2005; Górniak 2000).

2.2 Stream Discharge and Chemistry, Weather Data

The hydrochemical research was conducted between 19th and 31st of January 2009. Stream water stage was measured every 30 min, with a use of CTD Diver measuring device. These data were used for discharge calculations based on the rating curve, made for various stage heights encompassing the recorded stage during the whole monitoring period. Water samples for laboratory analyses were taken twice a day. Mixed water samples were collected from the center of the stream and transported directly to the laboratory.

Air temperature, sum of precipitations and solar energy were recorded electronically at 30 min intervals with a use of weather station Davis VantagePro2 installed within the catchment.

2.3 Water Sampling and Analysis

Groundwater and soil solution samples were taken from four plots situated along transverse transect through the valley. They represented different morphological and hydrological settings. Three plots: 1, 3 and 4 were located on arable fields: on the higher part of planar hillslope (1), hillslope-bench on the stream margin (3) and geomorphic hollow within hillslope (4), while plot 2 was situated on grassland in the waterlogged depression at the valley's bottom (Fig. 1). On each plot a nest of two wells was installed. The wells were cored to the depth of 0.6–2.0 m at which impeding clay was intersected. In one well a groundwater depth was recorded every 30 min using D-diver device (Schlumberger Ltd.). The second well was used for water sampling for chemical analyses.

Soil solution from A-horizons of soils (depth 0–0.2 m) was sampled using Eijkelkamp low vacuum soil moisture samplers with ceramic cup (12 samples). "Grab" snow samples and overland flow were sampled during active melt from five locations spread within the catchment (14 and 12 samples, respectively).

Water samples were collected in polyethylene bottles, transported directly to the laboratory and analysed for a large number of chemical variables. Those considered in this study are EC (Slandi SC 300), pH (HACH EC 10), Ca^{2+}, Mg^{2+} (AAS GBC Avanta), NH_4^+, SO_4^{2-}, NO_3^-, Cl^-, SiO_3^{2-}, PO_4^{3-} (Slandi LC 205 and Shimadzu UV–VIS 1800) and dissolved organic carbon DOC (Shimadzu TOC-V). DOC analyses were performed in the laboratory of Department of Hydrobiology University of Bialystok. Loads of solutes in stream outflow were calculated by:

$$l_k(j) = \sum_{i=1}^{k} \Delta t_i [c_i(j)Q_i + c_{i+1}(j)Q_{i+1}]/2, \qquad (1)$$

where $l_k(j)$, cumulative load component (j) in the time interval (k); Δt_i, time interval (i) between the measurements; $c_i(j)$, instantaneous concentration of component (j); Q_i, instantaneous flow rate (House et al. 2001).

2.4 Calculations and Statistical Analysis

To explore the contribution of each potential hydrological end-member to the stream discharge and chemistry as they vary in time throughout the course of the snowmelt high flow event, End-Member Mixing Analysis (EMMA; Christophersen and Hooper 1992; Burns et al. 2001) was performed on data representing river water sampled between 21 and 27 January 2009 and median chemistries of potential end-members measured for snowmelt conditions.

The potential end-members included: deeper groundwater contributing to stream base flow (BASE; for this end-member we used stream base flow chemistry recorded in pre-melt period), shallow groundwater beneath cropland (SGW) and soil solution (SS) collected in near stream locations, groundwater of near stream wetlands (WGW) and overland flow (OF). Moreover, we incorporated into analysis a virtual end-member, representing an average chemistry of saturated mineral soils, situated in the near-stream location (plot 3; PO). Ion composition of PO was calculated as median values of all data for SGW and SS form plot 3. Such approach was in our opinion justified because of close similarity of chemical composition between both components under saturation conditions during snowmelt.

EMMA analysis was preceded by initial source characteristic ("end-member") conducted on two dimensional mixing diagrams, in which Cl^-, Ca^{2+} and Mg^{2+} were used as conservative tracers. Streamflow solute concentrations were standardized and a correlation matrix was developed. A principal-components analysis (PCA; Statgraphics software) was performed on the correlation matrix. A model was selected that accounted for the greatest amount of variability with two principal components, implying a three end-member model. The EMMA model was then used to calculate the proportion of stream water derived from each of the developed end members for each streamwater sample (Burns et al. 2001).

3 Results

3.1 Weather, Streamflow Discharge and Groundwater Elevations

In the first days of January 2009 a prolonged period of freezing temperatures ($<-20°C$ at nights) led to development of soil frost whose depth reached up to 15–20 cm. After heavy snowfall in the second decade of January, the frozen soil was covered with snow layer of average thickness of 8–10 cm and water equivalent of 19–25 mm. Retention of water in snow cover and soil frost limited water circulation in the catchment. Stream exhibited very low flows of ~ 0.25–0.30 mm day^{-1} sustained mainly by deep groundwater discharging from sands and gravels of kame hills. During a day, the slight rise of temperature above zero coupled with intense sunlight caused local snowmelt, which resulted in short increase of flow in afternoon hours (Fig. 2). On the 22 of January, the rise of temperature above $2°C$ together with rain-on-snow event caused the first spate, which lasted to the morning hours on 23 of January. The next two even greater streamflow peak discharges appeared after rainfall on 24 and 25 January. By 25 January most of the catchment was snow-free. From 26 of January the stream discharge decreased gradually and by about 28 of January it reached the pre-melt value of 0.25–0.30 mm day^{-1} (Fig. 2).

Warming up caused gradual defrost of soil. After 26 of January on slopes of south and west exposition, the thickness of defrosted soil layer reached 5–10 cm. There was a 2–3 cm ice layer below. The total decay of soil ice was observed only locally at lower hillslopes and geomorphic hollows. Thawing on north slopes proceeded very slowly; in many areas the depth of de-frozen soil did not exceeded 2–4 cm and this state remained to the end of January.

The changes in groundwater level measured in wells at hillslope sites and in valley bottom did not follow simple or uniform patterns (Fig. 3). At the beginning of January groundwater was not present in hillslope settings. Low laying

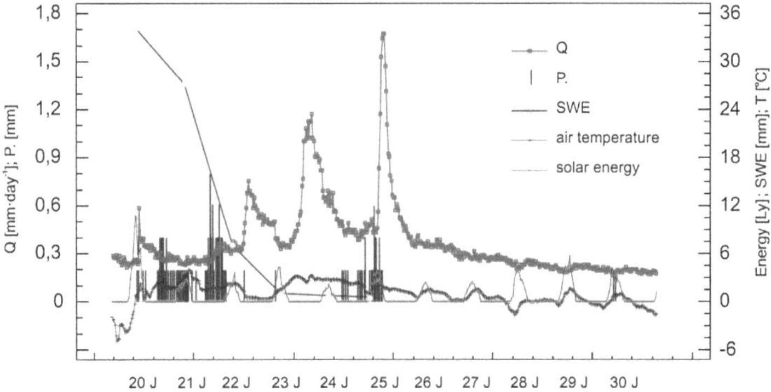

Fig. 2 Stream discharge Q and weather conditions during snowmelt high flow event between 19 and 31 January 2009; *P* rainfall, *SWE* snow water equivalent, *T* air temperature, *E* solar energy

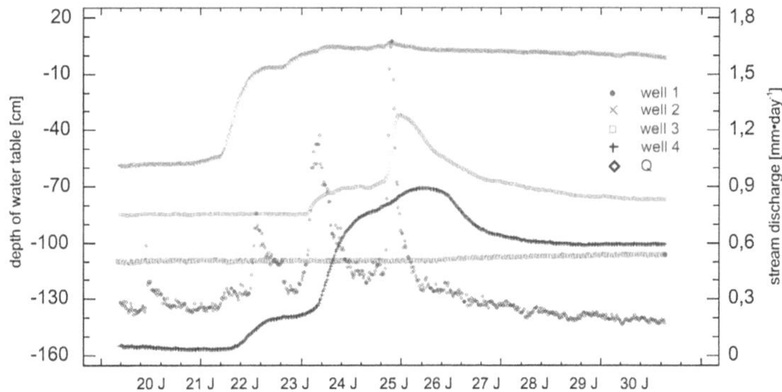

Fig. 3 Stream discharge and groundwater dynamics on the planar hillslope (well 1), grassland in the waterlogged depression at the valley's bottom (well 2), hillslope-bench on the stream margin (well 3) and geomorphic hollow within hillslope (well 4); groundwater depth measured from the ground surface

groundwater level (60 cm below land surface) was found for the geomorphic depression in the valley bottom (plot 2). Water level in the depression began to rise on about 21 of January during the first diurnal melt water input; after 22 January it reached quickly the ground surface and remained high without any significant changes to the end of the month.

At the hillslope-bench in near-stream location (well 3; thickness of loamy sands <70–80 cm) the water table appeared and started to rise on 24 of January and reached a maximum 43 h later after a heavy rain event. Water table began to recede just after peak stream discharge and returned very quickly to an approximately pre-melt level. The similar pattern of groundwater fluctuation was found for the well 4 located in a flat hollow within a midslope position. The total

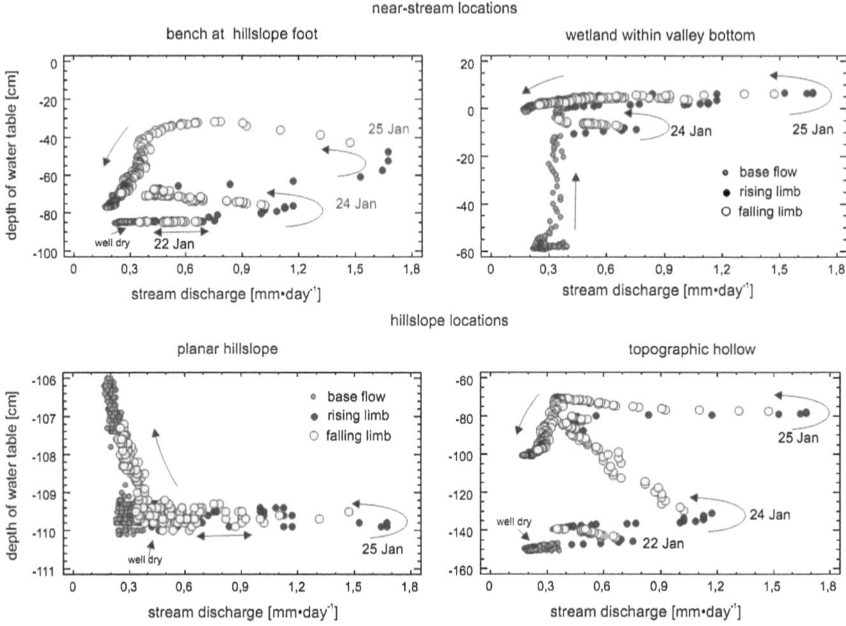

Fig. 4 Relationships between stream discharge Q and groundwater level

magnitude of water table rise was, however, considerably greater than that observed in lower slope settings. Moreover, after reaching its peak level groundwater table dropped but did not recede to its pre-melt state.

No reaction for hydrological changes in the catchment was observed at planar hillsope site (well 1, thickness of aquifer 100–120 cm). Thin, 1 cm water layer appeared above the impermeable till on the 26 of January. Its thickness rose insignificantly and on the last day of the month it reached only 4 cm (Fig. 3).

During the high flow period, linkage between groundwater level and stream flow (Q) was very weak. In general, during the first and second peak discharges, when ground was frozen or partly thawed, there was no free groundwater table on three of four investigated plots. As snowmelt progressed, groundwater level increased through the stream hydrograph recessions. Water table remained steady during the third peak discharge and rose again on the falling limb of hydrograph. At most sites there were a marked counterclockwise hysteresis in the groundwater level—stream discharge relation, what indicate that recharge of groundwater lagged the increase in streamflow (Fig. 4).

Outflow through till drain network appeared in the catchment on 28 of January. It was observed in one out of two monitored mouths of drain pipes and amounted to only 0.1–0.2 $dm^3 s^{-1}$, which indicates a slight infiltration of snowmelt water and a small recharge of soil water storage.

3.2 Solute Chemistry of Potential End-Members

Chemical composition of snow, overland flow, soil solution, and groundwater differed significantly (Table 1). The highest electrical conductivity EC values were observed in SS (900 $\mu S\ cm^{-1}$) and shallow groundwater on hillslopes beneath cropland (SGW 652 $\mu S\ cm^{-1}$), while the EC of water bound in snow cover (21 $\mu S\ cm^{-1}$) and of overland flow (174 $\mu S\ cm^{-1}$) was lowest. EC of groundwater under depression occupied by peatland WGW (well 2) reached 419 $\mu S\ cm^{-1}$. Concentration of most analyzed solutes changed in a similar way as EC. Soil solution and shallow groundwater under cropland had the highest concentration of NO_3^-, NH_4^+, Cl^-, DOC, Ca^{2+} and Mg^{2+}, while SiO_3^{2-} reached the highest concentration in deeper groundwater supporting stream base flow (BASE). Groundwater beneath peatland contained a very small amount of mineral forms of nitrogen, orthophosphate and silica, but it was high in sulfates.

3.3 Dynamics of Stream Chemistry

In general, an increase in discharge led to a decrease in specific conductance (EC; $r = -0.725$, $p < 0.01$) and concentrations of Ca^{2+} ($r = -0.678$), Mg^{2+} ($r = -0.713$) Cl^- ($r = -0.638$), Si ($r = -0.568$, $p < 0.05$) and SO_4^{2-} ($r = -0.661$), which may be linked to dilution of these ions that are derived mainly from groundwater and soil sources by near surface runoff under melt or stormflow conditions. From base flow to peak flow 3, EC decreased by a factor of 2. In contrast, positive correlations with flow were demonstrated by PO_4^{3-} ($r = 0.618, p < 0.01$). The highest concentration of PO_4^{3-}, reaching up to 1.5 mg dm^{-3}, occurred just after or during the peak flow.

The pH, NO_3^-, NH_4^+ and DOC showed no relationships with river discharge. A rise in stream discharge could cause an increase in pH and DOC concentration, as it happened on 22 and 24 of January, or their significant dilution, as it was observed on the 25 of January. NO_3^- concentrations increased through the rising limb, reaching a maximum during the recession stage. The concentration of nitrates increased steadily during the whole period of observation. In the middle of January it amounted to 5 mg dm^{-3}, while at the end of the month concentration reached 20 mg dm^{-3}, despite very similar flow conditions.

3.4 EMMA Results

Stream Ca^{2+}, Mg^{2+}, Cl^- and H^+ concentrations were used in a principal components analysis (PCA) after Christophersen and Hooper (1992). Two principal components were retained, which together explain nearly 80% of total variance,

Table 1 Chemistry of potential end-members contributing to the stream outflow during high flow between January 21 and 27 of 2009 (standard deviation in brackets)

End-member	EC $(\mu S\ cm^{-1})$	pH	NO_3^- $(mg\ dm^{-3})$	NH_4^+ $(mg\ dm^{-3})$	PO_4^{3-} $(mg\ dm^{-3})$	Ca^{2+} $(mg\ dm^{-3})$	Mg^{2+} $(mg\ dm^{-3})$	Cl^- $(mg\ dm^{-3})$	SO_4^{2-} $(mg\ dm^{-3})$	SiO_3^{2-} $(mg\ dm^{-3})$	DOC $(mg\ dm^{-3})$
S $n = 14$	21 (7)	5.05 (0.47)	2.6 (0.7)	0.71 (0.21)	0.20 (0.08)	10.0 (0.1)	2.0 (1.1)	6.2 (1.3)	3.8 (1.9)	0.1 (0.01)	1.2 (0.1)
OF $n = 12$	174 (73)	7.22 (0.27)	10.6 (10.9)	0.70 (0.19)	1.15 (1.54)	19.6 (4.7)	6.9 (2.9)	11.4 (5.5)	7.6 (5.9)	3.1 (0.8)	22.9 (4.0)
SS $n = 6$	900 (278)	5.95 (0.14)	85.0 (53.4)	3.88 (2.28)	0.26 (0.05)	118.6 (43.2)	21.0 (6.9)	80.0 (31.6)	59.3 (24.4)	4.1 (0.8)	88.6 (19.9)
SGW $n = 6$	652 (26)	6.90 (0.35)	80.2 (17.9)	0.58 (0.13)	0.17 (0.07)	85.5 (12.3)	13.6 (2.1)	24.9 (6.1)	73.3 (5.0)	2.6 (0.3)	51.9 (20.0)
PO $n = 6$	776 (176)	6.43 (0.56)	81.2 (36.6)	2.23 (2.33)	0.21 (0.08)	102.0 (26.4)	17.3 (4.8)	56.0 (27.2)	66.3 (18.0)	3.3 (0.9)	67.6 (26.8)
WGW $n = 4$	419 (8)	7.10 (0.30)	2.6 (2.1)	0.63 (0.32)	0.10 (0.01)	64.9 (3.5)	12.4 (1.4)	4.4 (3.4)	79.5 (3.7)	1.8 (0.3)	46.7 (14.6)
BASE $n = 3$	690 (40)	7.37 (0.04)	4.0 (0.6)	0.27 (0.06)	0.02 (0.01)	110.0 (3.9)	21.1 (1.3)	22.7 (0.6)	73.7 (7.0)	4.5 (0.7)	40.7 (4.1)

S, water bound in snow cover; OF, overland flow; SS, soil water collected in near-stream locations; SGW, shallow groundwater beneath arable lands in near-stream locations; PO, mixture of soil and groundwater beneath cropland in near-stream locations (averaged values from suction lysimetres and wells); WGW, groundwater of near stream wetlands; BASE, deeper groundwater contributing to stream base flow; n, number of samples

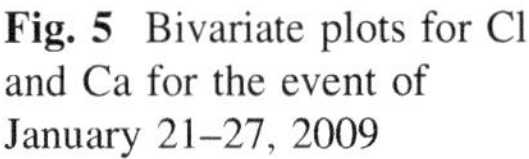

Fig. 5 Bivariate plots for Cl and Ca for the event of January 21–27, 2009

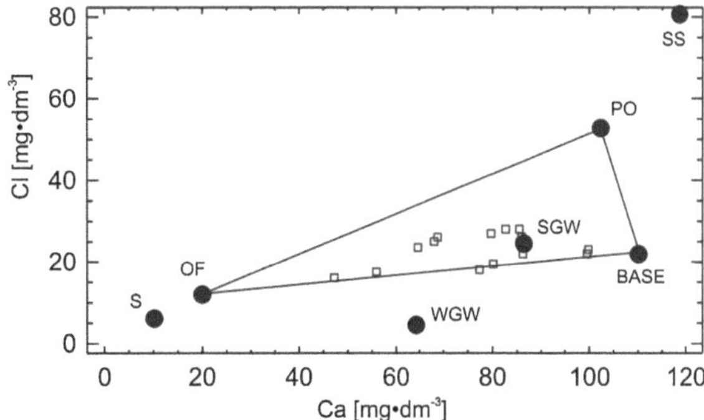

implying that at least three potential sources (end members) may have an influence on the variation of river chemistry.

Cl–Ca bivariate plots and EMMA mixing diagrams show four likely end-members: SGW, PO, OF and BASE. However, streamwater was enclosed by OF, PO and BASE, therefore, they were assumed as main sources controlling stream chemistry during snowmelt flood (Fig. 5).

The EMMA model was evaluated by comparing the model-predicted concentrations for SO_4^{2-}, PO_4^{3-}, DOC and NO_3^- against observed stream concentrations. The R^2 values of linear regression model for sulphate and phosphate varied between 61 and 72%, indicating that the selected EMMA model was a good predictor of stream solute concentrations. The ability of EMMA for prediction of temporal pattern of DOC and NO_3^- concentration was much weaker, with R^2 amounted 8 and 20%, respectively, suggesting non-conservative mixing of these compounds.

EMMA-derived hydrograph separations showed an important role of overland flow (OF) in contributing to the stream discharge. Percentage of OF was highest during the peak flow, when it amounted up to 70% of Q, but its occurrence on the level of 5–25% was observed during whole period between 21 and 27 of January. Contribution of deeper groundwater BASE was highest before the start of the snowmelt, declined very quickly during peak discharges and rose again during hydrograph recessions. Input of PO increased on the hydrograph rising limbs and reached a maximum after the last peak discharge. Its biggest share in total stream Q amounted to 50% and occurred between 26 and 27 of January, when infiltrating snowmelt and rainfall filled storage on the hillslope (Fig. 6).

The observed and calculated with EMMA model loads of solutes exhibited a great similarity (Table 2). Usually, differences did not exceed 6%; only in case of silica and DOC the inconsistency was bigger and amounted to 21 and 24%, respectively.

Deeper groundwater feeding the base flow (BASE) together with overland flow (OF) were the most important components of catchment runoff. The surface runoff was the leading contributor of orthophosphate and was also significant in the migration of NH_4^+, NO_3^-, Si and DOC. BASE was main source of base cations.

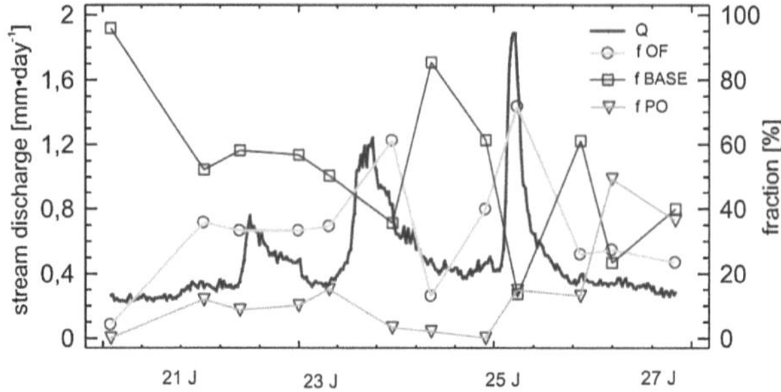

Fig. 6 End-members contribution to stream discharge between January 21 and 27, 2009

Table 2 Observed and EMMA model-predicted loads of solutes in stream outflow between January 21 and 27 of 2009

Load	Q (m^3)	NO$_3^-$ (kg)	NH$_4^+$ (kg)	PO$_4^{3-}$ (kg)	Ca^{2+} (kg)	Mg^{2+} (kg)	Cl$^-$ (kg)	SO$_4^{2-}$ (kg)	SiO$_3^{2-}$ (kg)	DOC (kg)
Observed	5695.7	88	3.3	3.1	391	86	123	285	17	162
Predicted	5784.4	93	4.1	2.9	415	85	126	270	22	212
Percentage share										
OF	41	27	40	93	11	19	22	7	34	26
PO	13	61	42	5	18	15	30	20	11	23
BASE	46	11	17	2	71	66	48	73	55	51

Saturated mineral soils placed at the foot of hillslope, in the close neighborhood of watercourse (PO) were the most important source of NO$_3^-$ and NH$_4^+$.

4 Discussion

Numerous authors have well documented the importance of hydrological events in affecting both the hydrological and chemical response of surface waters (Banaszuk 2004; Inamdar et al. 2004; Inamdar and Mitchell 2006; Mitchell et al. 2006; Petry et al. 2002). It was found that the largest export of solutes from watersheds occurred during large runoff events induced by intense rainfall or snowmelt. Most commonly, the enhanced fluxes of solutes observed at that time in a catchment system were attributed to the displacement of the so-called old or pre-event water, rich in dissolved compounds, by a front of infiltrating event water, thus soil layers or groundwaters near the surface were identified as major sources of NO$_3^-$, SO$_4^{2-}$ and dissolved organic carbon (DOC). Event water was found to be a relatively minor contributor to stream chemistry during storms and snowmelt events (Burns et al. 2001; Kendall and McDonnell 1998; Laudon et al. 2004). Its prominent role was documented only in few studies (e.g., Kendall et al. 1999).

In studied catchment, the stream discharge and chemical composition of the streamwater during snowmelt induced high flow event were controlled by the shallow flow path components. From the very early beginning of snowmelt, when temperature-induced snowmelt coupled with rainfall began to produce an overland flow that caused the hydrograph to rise, overland and rill flows seemed to be the main mechanism of runoff generation because of the low permeability of the frozen ground. In such conditions meltwater may become Hortonian overland flow on top of a soil layer (Laudon et al. 2004).

Despite dynamic changes in stream outflow, in the first days of snowmelt we did not find any groundwater response to snowmelt input on the hillslope locations (Fig. 4). Groundwater level started to grow only on 24 January but its intensity differed substantially depending on landscape position and surficial and subsurface impermeable till layer topography.

Enhanced infiltration occurred under small depressions, geomorphic hollows and hillslope benches, ponded by snowmelt water (well 3 and 4). In these places infiltration might progress via vertical unfrozen conduits that bypass the soil matrix, rapidly transporting percolates downwards. Depression-focused recharge of groundwater by snowmelt water infiltration through large pores, and cracks in frozen soils has been documented, i.e., by Baker and Spaans (1997) and Sharratt (2001). Hydrometric and chemical evidence imply that shallow groundwater and soil water originated in hollows especially that located in near-stream positions, may be an important contributor to the stream discharge and chemistry in recessive stage of high flow events.

By contrast, vertical water movement was insignificant on planar hillslopes, where soil frost reduced the infiltration, due mainly to the blocking effect of the ice. During the whole period of snowmelt, in well 1 located on upper hillslope, the groundwater zone rose by a few centimeters only, reaching the total depth of 4 cm above the underlying loam. Hydrometric observations suggest that during studied snowmelt event subsurface hydrology of such element of landscape was largely decoupled from the stream system. There was neither substantial groundwater movement in the stream direction, nor water supply to the topographical depression at the foot of hillslope (well 2). Thus, planar hillslopes constituted transition zones for surface runoff only.

The replenishment of groundwater storage occurred with the great intensity in the depression in valley bottom filled by shallow peat and occupied by permanent grassland. At this place (well 2), at the first days of event the water table rose up to the ground surface (with a shallow ponding) and remained high for days after peak streamflow. As depression bordered the described above planar hillslope, which created its only upslope contributing area, it was unlikely that observed water saturation might be attributed to the groundwater recharge. Increase in groundwater level was most probably caused by infiltration of melt water. Water table in wetland showed no relationship to the dynamics of stream discharge. We relate this lack of coincidence to the geomorphological settings of the depression. Wetland was separated from streambed by kind of underground "embankment"— a narrow strip of sediments with low hydraulic conductivity (sandy loams and peaty sediments with high silt admixture) that limited subsurface hydrologic connections. Moreover, streambanks were higher by 10–30 cm compared to wetland surface, in places they have been elevated during stream regulation works, which disrupted a free transit of surface outflow. Some water flux toward the stream may occur via lateral preferential flowpaths, e.g., abundant mole tunnels. However, hydrometric data rather excluded an active and a rapid groundwater movement from the wetland depression to the stream. This little potential for water supply to the watercourse has been also confirmed by results of EMMA analysis.

To recapitulate, presented evidences suggest that during investigated snowmelt event, despite a substantial recharge of groundwater storage, only a small percentage of the landscape might be a source of subsurface flow of solutes that contributed to the streamwater discharge and chemistry. They were mainly saturated areas well-connected to the drainage network: hollows and sideslope benches along the stream margins. In our catchment shallow groundwater and soil water discharged from those locations were responsible for pronounced ($>60\%$) export of nitrates. Isolated saturated areas, like a wetland in depression on the valley bottom, will rather serve as sites for the loss of NO_3^- due to dissimilatory nitrate reduction before this water reaches the stream (Inamdar and Mitchell 2006). Groundwater in wetland was found to have the lowest NO_3^- concentration, which did not exceeded 4.5 mg dm^{-3}.

The dominant shallow hydrological flowpaths: overland flow above the frozen soil layer and shallow ground- and surface waters generated translocation of 88% of nitrate loads and up to 98% of orthophosphate from the catchment. In the migration of phosphorus, the surface flow played a key role; it was also important in translocation of silica and ammonium washed out from the surface of frozen or partly thawed soils.

The high concentration of dissolved compounds (primarily nitrate) in the river outflow leads us to the supposition that during the period of most intensive chemical denudation in NE Poland, fluxes of agricultural contaminants either bypass potential structures or buffers that could constrain their impact on the freshwater ecosystems or existing buffers are ineffective in removing contaminants moving along shallow hydrological pathways from diffuse surface and near surface sources. Therefore, the short period of snowmelt in early spring may be perceived as critical from the river water quality perspective.

In early spring, high export of NO_3^- along surface and shallow subsurface hydrological pathways may be little affected by the vegetation of the widely promoted buffer strips because at this time of the year the vegetation is still in a dormant phase. Tree and grassland strips located along the Narew Valley may, however, provide an important barrier for sediment-bound nutrients (Banaszuk 2007).

5 Conclusions

Streamwater discharge and chemistry during snowmelt were largely controlled by the shallow flow path components: overland flow and shallow ground- and surface waters from the near-stream locations. Solutes transported along with overland flow had the most pronounced impact on stream chemistry during peak discharge, when this runoff component constituted up 70% of stream discharge. High surface runoff contributions produced a pronounced rise and maximum in streamflow PO_4^{3-} concentration.

We found that during investigated snowmelt event only a small percentage of the landscape generated a subsurface water flow that contributed to the stream discharge. They were mainly saturated areas well-connected to the watercourse; water flux discharging from those locations led to pronounced export of nitrates.

Hydrometric and chemical evidence imply that during snowmelt event of 2009 planar hillslopes were hydrologically decoupled from the stream system. There was neither substantial recharge of hillslope storage nor groundwater movement in the stream direction. We stated that planar hillslopes constituted transition zones for surface runoff only.

The high concentration of solutes in the river outflow suggests that during snowmelt, either fluxes of agricultural contaminants bypassed potential buffers or that existing buffers were ineffective in removing the contaminants that moved along shallow flow paths. Thus, the short period of snowmelt flood may be perceived as critical from the river water quality perspective.

Acknowledgments The authors would like to thank dr E. Jekatierynczuk-Rudczyk of the University in Bialystok for performing DOC analyses.This research was funded by Technical University of Bialystok (grant S/IIS/21/08).

References

Atlas klimatu Polski (2005) In: Lorenc H (ed) Polish Climate Atlas. IMiGW, Warszawa, p 116 (in Polish)

Baker JM, Spaans EJA (1997) Mechanics of meltwater movement above and within frozen soil. In: Iskander IK, Wright EA, Radke JK, Sharratt BS, Groenvelt PH, Hinzman LD (eds) International symposium on physics, chemistry and ecology of seasonally frozen soils. US Army Cold Regions Research and Engineering Laboratory, Fairbanks, pp 31–36

Banaszuk P (2004) Identyfikacja procesów kształtujących skład chemiczny małego cieku w krajobrazie rolniczym na podstawie analizy czynnikowej. Woda Środ Obsz Wiej 4, 1(10):103–116

Banaszuk P (2007) Wodna migracja składników rozpuszczonych do wód powierzchniowych w zlewni górnej Narwi. Wydaw P Biał, Białystok, p 182

Burns DA, McDonnell JJ, Hooper RP, Peters NE, Freer JE, Kendall C, Beven K (2001) Quantifying contributions to storm runoff through end-member mixing analysis and hydrologic measurements at the Panola Mountain Research Watershed (Georgia, USA). Hydrol Process 15:1903–1924

Campbell JL, Mitchell MJ, Groffman PM, Christenson LM, Hardy JP (2005) Winter in northeastern North America: a critical period for ecological processes. Front Ecol Environ 3(6):314–322

Christophersen N, Hooper RP (1992) Multivariate analysis of stream water chemical data: the use of principal component analysis for the end-member mixing problem. Water Resour Res 28:99–107

Christophersen N, Neal C (1990) Linking hydrological, geochemical and soil chemical processes on the catchment scale: an interplay between modeling and fieldwork. Water Resour Res 26:3077–3086

Creed IF, Band LE, Foster NW, Morrisom IK, Nicolson JA, Semkin RS, Jeffries DS (1996) Regulation of nitrate-N release from temperate forests: a test of the N flushing hypothesis. Water Resour Res 32(11):3337–3354

Górniak A (2000) Klimat województwa podlaskiego. IMGW, Białystok, p 119

Haag D, Kaupenjohann M (2001) Landscape fate of nitrate fluxes and emissions in Central Europe. A critical review of concepts, data, and models for transport and retention. Agric Ecosyst Environ 86, 1:1–21

Hart G (1963) Snow and frost conditions in New Hampshire, under hardwoods and pines and in the open. J For 61, 4:287–289

House WA, Leach DV, Armitage PD (2001) Study of dissolved silicon and nitrate dynamics in a freshwater stream. Water Res 35(11):2749–2757

Inamdar SP, Mitchell MJ (2006) Hydrologic and topographics controls on storm-event exports of dissolved organic carbon (DOC) and nitrate across catchment scales. Water Resour Res, vol 42, p W03421, doi: 10.1029/2005WR004212

Inamdar SP, Christopher SF, Mitchell MJ (2004) Export mechanisms for dissolved organic carbon and nitrate during storm events in a glaciated forested catchment in New York, USA. Hydrol Process 18:2651–2661

Kendall C, McDonnell JJ (1998) Isotope tracer in catchment hydrology. Elsevier, New York, p 870

Kendall KA, Shanley JB, McDonnell JJ (1999) A hydrometric and geochemical approach to test the transmissivity feedback hypothesis during snowmelt. J Hydrol 219:188–205

Laudon H, Seibert J, Köhler S, Bishop K (2004) Hydrological flow paths during snowmelt: congruence between hydrometric measurements and oxygen 18 in meltwater, soil water, and runoff. Water Resour Res, vol 40, p W03102, doi: 10.1029/2003WR002455

Mitchell MJ, Driscoll CT, Kahl JS, Likens G, Murdoch P, Pardo L (1996) Climatic control of nitrate loss from forested watersheds in the northeast United States. Environ Sci Technol 30:2609–2612

Mitchell MJ, Piatek KB, Christopher S, Mayer B, Kendall C, Mchale P (2006) Solute sources in stream water during consecutive fall storms in a northern hardwood forest watershed: a combined hydrological, chemical and isotopic approach. Biogeochemistry 78:217–246

Petry J, Soulsby C, Malcolm IA, Youngson AF (2002) Hydrological controls on nutrient concentrations and fluxes in agricultural catchments. Sci Total Environ 294:95–110

Piatek KB, Mitchell MJ, Silva SR, Kendall C (2005) Sources of nitrate in snowmelt discharge: evidence from water chemistry and stable isotopes of nitrate. Water Air Soil Pollut 165:13–35

Pionke HB, Gburek WJ, Sharpley AN (2000) Critical source area controls on water quality in an agricultural watershed located in the Chesapeake Bay. Ecol Eng 14:325–335

Shanley JB, Chalmers A (1999) The effect of frozen soil on snowmelt runoff at Sleepers River. Vt Hydrol Process 13:1843–1857

Sharratt BS (2001) Groundwater recharge during spring thaw in the Prairie Pothole Region via large, unfrozen preferential pathways. II. International symposium on preferential flow, Am Soc Agric Eng, Honolulu, pp 49–52

Sickman JO, Leydecker A, Chang CCY, Kendall C, Melack JM, Lucero DM, Schimel J (2003) Mechanisms underlying export of N from high-elevation catchments during seasonal transitions. Biogeochemistry 64:1–24

Smart RP, Soulsby C, Neal C, Wade A, Cresser MS, Billet MF, Langan SJ, Edwards AC, Jarvie HP, Owen R (1998) Factors regulating the spatial and temporal distribution of solute concentrations in a major river system in NE Scotland. Sci Total Environ 221:93–110

Soulsby C, Gibbins C, Wade AJ, Smart R, Helliwell R (2002) Water quality in the Scottish uplands: a hydrological perspective on catchment hydrochemistry. Sci Total Environ 294:73–94

Szpikowski J, Kostrzewski A, Mazurek M, Smolska E, Stach A (2008) Współczesne procesy kształtujące rzeźbę stoków. In: Starkel L, Kostrzewski A, Kotarba A, Krzemień K (eds) Współczesne przemiany rzeźby Polski. IG i GP UJ, Kraków, pp 283–291

Wang FL, Bettany JR (1993) Influence of freeze-thaw and flooding on the loss of organic carbon and carbon dioxide from soil. J Environ Qual 22:709–714

Wang FL, Bettany JR (1994) Organic and inorganic nitrogen leaching from incubated soils subjected to freeze-thaw and flooding conditions. Can J Soil Sci 74:201–206

Modelling of Groundwater Fluctuations in the Area Between the Narew and Suprasl Rivers

Waldemar Mioduszewski, Erik P. Querner and Justyna Bielecka

Abstract The main goal of this paper is to explore by hydrological modelling the groundwater system on both sides of the watershed boundary between the rivers Narew and Supraśl, North-East Poland and the influence of two nearby peat mines. An integrated model of the area was set up using the SIMGRO program. The aim of the model simulations was to estimate the historic hydrological situation of the area. The current groundwater levels were simulated as well as scenarios were investigated, which were based on blocking all drainage canals in the area of the two peat mines. The research proves that the watershed boundary between the Narew and the Suprasl River corresponds to the division of groundwater basins. Significant influences of the peat mines on groundwater levels are noticeable only in the close surroundings of the mines.

1 Introduction

The area of the study, which is located in the north-eastern part of Poland, covers the surroundings of the watershed division between the Narew and the Suprasl Rivers. In the past, hydrological system of the place was changed

W. Mioduszewski (✉) · J. Bielecka
Department of Water Resources, Institute of Technology and Life Science,
Falenty, Al. Hrabska 3, 05-090 Raszyn, Poland
e-mail: w.mioduszewski@itep.edu.pl

E. P. Querner
ALTERRA, Centre for Water and Climate, P.O. Box 47, 6700 AA Wageningen,
The Netherlands

D. Mirosław-Świątek and T. Okruszko (eds.), *Modelling of Hydrological Processes in the Narew Catchment*, Geoplanet: Earth and Planetary Sciences,
DOI: 10.1007/978-3-642-19059-9_5, © Springer-Verlag Berlin Heidelberg 2011

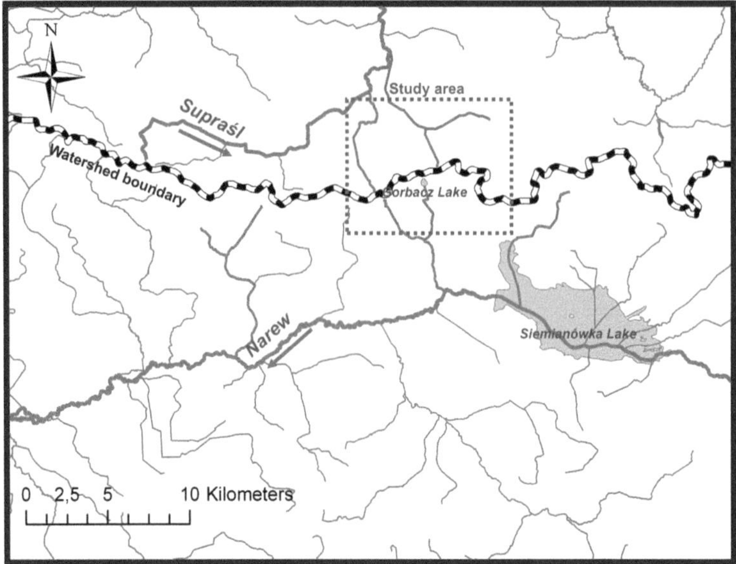

Fig. 1 Localization of the study area

considerably in order to adapt land for extensive agriculture. Moreover for excavation of peat, a lot of canals and small drainage ditches were built. Two peat mines were built, one called Rabinowka in the Suprasl catchments, and the other—Imszar in the Narew one. In the area the small and shallow lakes Gorbacz (Banaszuk et al. 1994) and Wiejki are located. Both have high ecological values and should be protected.

The research was focused on exploring the hydrological regime of the study area and the relation between the peat excavation and groundwater levels in the surroundings. Also is assessed the influence of the dimension of peat mine on groundwater levels and water levels in Gorbacz Lake. The purpose of the simulations was to estimate the previous hydrological conditions, before the peat mines were built and the drainage system was constructed. The study was carried out by means of hydrological modelling, which enables to set up a model of the area and simulate different scenarios. A specific hydrological problem has been presented in the paper. Two valuable natural wetlands are located on both sides of the watershed boundary at a small distance from each other. Next to the wetlands peat mines are located (Fig. 1). It is necessary to identify the impact of the two mines on the moisture conditions of the protected wetlands as well as on the location of the watershed boundary and its probable shift as a result of dewatering of the mine.

2 The Combined Surface and Groundwater Flow SIMGRO Model

In many practical applications, models are used as predictive tools, to evaluate various water management measures, policies or scenarios. The SIMGRO (SIMulation of GROundwater and surface water levels) groundwater model we applied to the peatlands has two objectives: systems analysis and prediction. It is a physically based model that simulates regional transient saturated groundwater flow, unsaturated flow, actual evapotranspiration, stream flow, groundwater and surface water levels as a response to rainfall, reference evapotranspiration, and groundwater abstraction. To model regional groundwater flow, as in SIMGRO, the system has to be schematised geographically, both horizontally and vertically. The horizontal schematisation allows different land uses and soils to be input per node, to make it possible to model spatial differences in evapotranspiration and moisture content in the unsaturated zone. For the saturated zone, various subsurface layers are considered (Fig. 2). For a comprehensive description of SIMGRO, including all the model parameters, readers are referred to Van Walsum et al. (2004) or Querner (1997).

The SIMGRO model is used within the GIS environment Arcview. Via the user interface AlterAqua, digital geographical information (soil map, land use, water courses, etc.) can be input into the model. The results of the modelling are analysed together with specific input parameters.

2.1 Groundwater Flow

In SIMGRO the finite element procedure is applied to approach the flow equation which describes transient groundwater flow in the saturated zone. A transmissivity is allocated to each node to account for the regional hydrogeology. A number of nodes make up a subcatchment, as shown in Fig. 2. The unsaturated zone is represented by means of two reservoirs: one for the root zone and one for the underlying substrate (Fig. 2). The calculation procedure is based on a pseudo-steady state approach, generally using time steps of up to 1 day. If the equilibrium moisture storage for the root zone is exceeded, the excess water will percolate towards the saturated zone. If the moisture storage is less than the equilibrium moisture storage, then water will flow upwards from the saturated zone (capillary rise). The depth of the phreatic surface is calculated from the water balance of the subsoil below the root zone, using a storage coefficient. The equilibrium moisture storage, capillary rise and storage coefficient are required as input data and are given for different depths to the groundwater. Evapotranspiration is a function of the crop and moisture content in the root zone. To calculate the actual evapotranspiration, it is necessary to input the measured values for net precipitation, and the potential evapotranspiration for a reference crop (grass) and woodland. The model derives the potential evapotranspiration for other crops or vegetation types

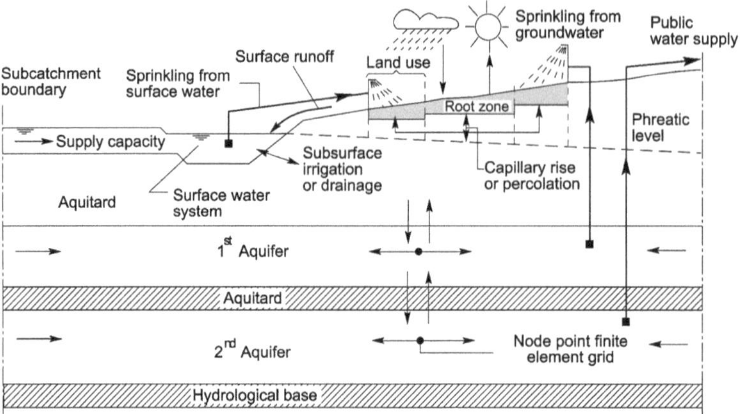

Fig. 2 Schematization of water management in the SIMGRO model. The main feature is the integration of the saturated zone, unsaturated zone and surface water systems within a subregion (Querner 1997)

from the values for the reference crop, by converting with known crop factors (Feddes 1987). Snow accumulation has been accounted for in the model: it is assumed that snow accumulation and melting is related to the daily average temperature. When the temperature is below 0°C, precipitation falls as snow and accumulates. At temperatures between 0 and 1°C, both precipitation and snow melt occur: it is assumed that during daylight hours the precipitation falls as rain, whereas precipitation falling during the night accumulates as snow (and the melt rate is 1.5 mm water per day). When the temperature is above 1°C, the snow melts at a rate of 3 mm/day per degree Celsius. Taking the winter season into consideration during the construction of the model is particularly important as the described area is situated in the region of Poland with the lowest temperatures in winter (January −4°C, February −3°C) and the longest period with frost (on average 100 days per year).

2.2 Surface Water Flow

The surface water system in peatlands usually consists of a natural river and a network of small water courses, lakes and pools. It is not feasible to explicitly account for all these water courses in a regional simulation model, yet the water levels in the smaller water courses are important for estimating the amount of drainage or subsurface irrigation, and the water flow in the major water courses is important for the flow routing. The solution is to model the surface water system as a network of reservoirs. The inflow into one reservoir may be the discharge from the various water courses, ditches and runoff. The outflow from one reservoir is the inflow to the next reservoir. The water level depends on surface water storage and on reservoir inflow and discharge. For each reservoir, input data are required on two relationships: "stage versus storage" and "stage versus discharge".

2.3 Drainage

Water courses are important for the interaction between surface water and groundwater. In the model, four different categories of ditches (related to its size) are used to simulate the drainage. It is assumed that three of the subsystems—ditches, tertiary watercourses and secondary watercourses—are primarily involved in the interaction between surface water and groundwater. A fourth system includes surface drainage to local depressions. The interaction between surface and groundwater is calculated for each drainage subsystem using a drainage resistance and the hydraulic head between groundwater and surface water (Ernst 1978).

2.4 Linkage of Groundwater and Surface Water Modules

As the groundwater part of the model reacts much more slowly to changes than the surface water part, each part has its own time step. As a result, the surface water module performs several time steps during one time step of the groundwater module. The groundwater level is assumed to remain constant during that time and the flow between groundwater and surface water accumulates using the updated surface water level. The next time the groundwater module is called up, the accumulated drainage or subsurface irrigation is used to calculate a new groundwater level.

3 Description of the Study Area and Schematization

The study area covers 8,192 ha and it was divided into 5,168 nodal points, which were characterized by geographic coordinates and surface levels. North, west and south borders of the area were designated along the catchments borders, whereas east border is the water course Rudniki (Fig. 3). The watershed boundary of Narew and Suprasl Rivers is situated in the middle of the modelled area.

The entire area is an example of a typical glacial landscape with a lot of glacial relics formed by moraines: kettles, valleys formed by melt water, kamers, lowland of sandr and small hills. Some of the bigger lowlands are filled by peat whereas smaller ones by mineral soils (Mioduszewski 2004). The biggest lowland is situated in the upper part of the catchment of the River Suprasl, and it is in the central

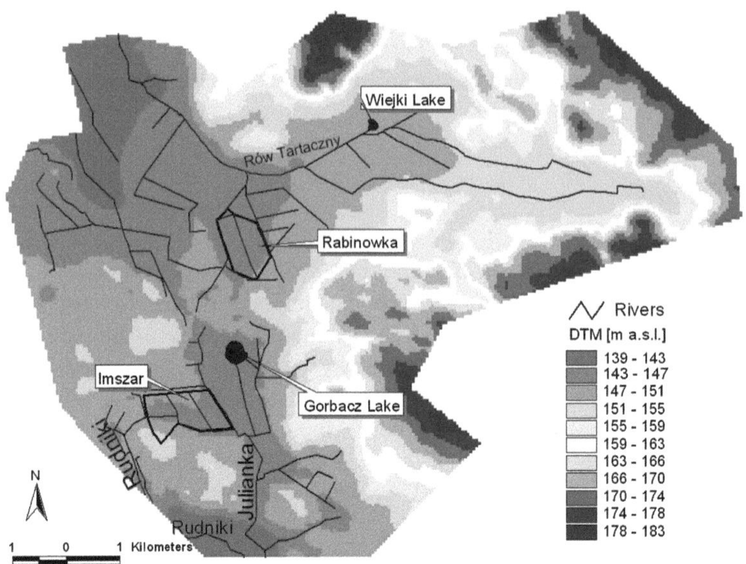

Fig. 3 Elevation of the study area

part of the modelled area. It is filled by peat soil with several sandy hills and surrounded by uplands, which consists mostly of sandy soils and in some small holes mud is present. Based on hydrological studies it is assumed that two geological strata of different thickness are present (Banaszuk 1995; Mioduszewski 2004). The first layer, with a thickness between 30 cm and 6 m, consists of peat, sand and mud, and its hydraulic conductivity is assumed to range from 0.01 m/day for peat up to 0.3 m/day for sand. Underneath the first strata there is a layer of sand 10–17 m deep, which has a hydraulic conductivity of 22 m/day. The clay layer beneath the sand is considered to be impermeable.

The peat territory of the lowland is a matrix of grasslands and deciduous forests. The area of both peat mines is covered by grass vegetation. The Gorbacz Lake banks are overgrown by mosses, reeds and willow vegetation and surrounded by sedges, willow and birch forest (Banaszuk 1995). Based on existing maps and field studies 7 types of land use were designated: grassland, arable land, deciduous forest, pine forest, fresh water, reeds, built-up area.

As mentioned above, the area belongs to catchments of two big rivers: Narew and Suprasl, however both of them are not present itself in the modelled area. Row Tartaczny, Julinaka and Row Rudniki are the bigger water courses which flow through the territory. Besides them also a number of small canals and drainage ditches are present. For the Simgro model, rivers and canals were divided into main water courses and drainage ditches. The main rivers and mines were described as linear streams with defined location of surface water. Smaller streams were modeled as a drainage subsystem (Querner 1997; Ernst 1978).

Meteorological data, as measured in the Institute of Meteorology observatory station in the village Biebrza, were used. The station is located about 20 km from the study area and it is the closest station that has full set of observation parameters that allow the calculation of evapotranspiration. The data consists of daily values of precipitation (mm) and air temperature (°C) for the simulated period 1996–2001. From the meteorological data potential evapotranspiration for reference, pine and deciduous forest was calculated.

The condition considered in every nodal point along the boundary of the model was of the type no flow. Based on the field observations (shallow drillings, water measurements in wells of farms) along the external borders of the area a steady level of the groundwater table has been assumed. Along the streams the existing surface water position (a reference scenario) and the variable water table position in the peat mine have been assumed.

4 Simulations and Results

For investigation of the hydrological system of the area two different scenarios were simulated. The first scenario shows the present hydrological situation of the area and is based on existing data thus it was used as the reference situation. It gives an overview of the hydrology of the area. This scenario was verified using the field measurements. Groundwater level measurements have been carried out in 10 piezometers located around the peat mine (6 piezometers near the Rabinówna mine and 4 piezometers near the ISZMAR mine). The difference between the calculated and measured groundwater table is ±0.2 m. The accuracy has been considered as sufficient as the purpose of the modeling was to find differences between the historic and present water table level.

The second scenario was created to obtain the natural hydrological regime of the area before excavation of peat started and is based on blocking all canals and smaller drainages ditches in the areas of two peat mines. It could give a general idea of the influence of the peat mines on the groundwater levels as well as surface levels in the Gorbacz Lake.

Groundwater levels calculated from the model simulation range from 139.00 m a. s. l. till 180.55 m a.s.l. (Fig. 4). The lowest values occur in the south-east and south-west parts of the area and higher are reached in the area of the western uplands. The watershed boundary between the catchment of the Suprasl River and catchment of the Narew River is clearly marked in the central part of the area and corresponds to the estimation in the field.

Blocking of every canal in the area of the two mines would cause significant rise of groundwater levels in the close neighborhoods of the mines. However neither blocking canals in Imaszar nor in Rabinowka would influence the surface water levels in the Gorbacz Lake. If this scenario is representative for the situation before excavation of peat in the area started, then the building of neither of the two mines could have influenced the water level in the Gorbacz Lake. The influence

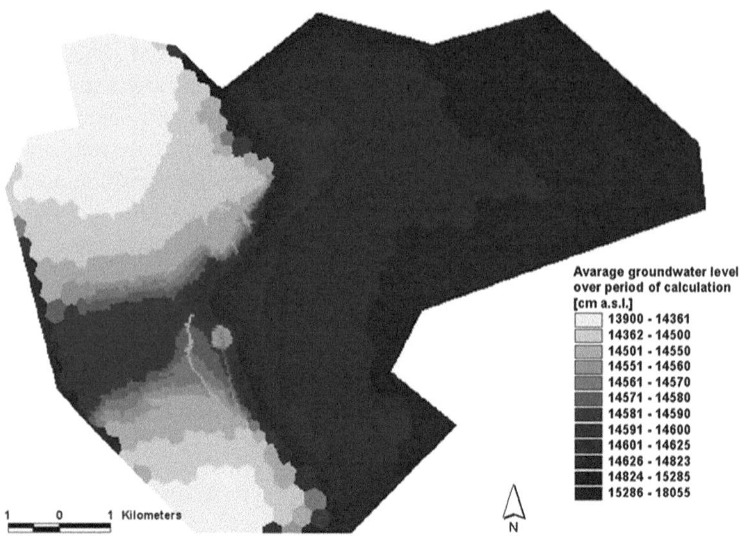

Fig. 4 Average groundwater levels of the study area as simulated by the model

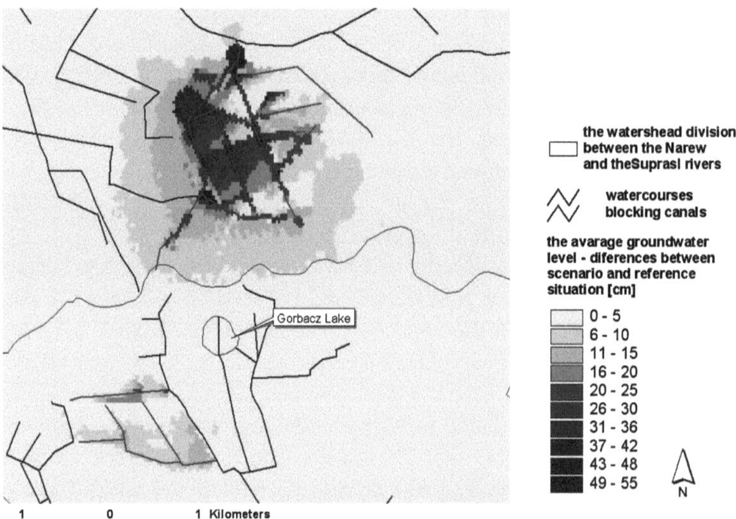

Fig. 5 The average groundwater levels—differences between scenario and reference situation

would have been local and not exceed the watershed boundary and there are no shifts in the position of the watershed boundary. Although Rabinowka mine could locally influence the Narew catchment, its influence would be present only in some nodal points and they would not be higher than 6 cm (Fig. 5). An area, where groundwater level would increase considerably with 6–10 cm, would be situated in the south and east surroundings of the Rabinowka mine. However, changes of groundwater levels higher than 20 cm would occur only in the area close to the blocked canals and higher than 40 cm merely in some particular nodal points. Blocking of canals and drainage ditches in the Imszar mine would cause the rise of groundwater level only along blocked canals and in some parts of the mine. Changes would be mostly between 6 and 10 cm and along canals 10–20 cm.

5 Conclusions

This preliminary study shows that the SIMGRO model can be used for predicting the influence of peat mines on groundwater levels. The surface watershed division between the Narew River and the Suprasl River corresponds with the border of the groundwater basins. There are no changes in location of the watershed boundary due to the raising of the water level. Model simulations show that both mines, Rabinowka and Imszar, are not influencing the water level in the Gorbacz Lake. Blocking of canals in the area of the two mines would cause an increase of groundwater levels in the surrounding. The influence of the Rabinowka mine is significantly larger than the influence of Imszar mine.

References

Banaszuk H (1995) Influence of business on soils and soil environment in the Imszar trough and the Gorbacz Reserve. In: Soil environment of the protected areas in Białostocczyzna, pp 59–66

Banaszuk H, Banaszuk P, Bartoszuk H, Kondratiuk P, Stepaniuk M (1994) Water dislocation in peat bog reserves and its influence on nature as described in "Gorbacz" reserve. In: Ekonomia I Środowisko, vol 1, no. 4, pp 115–133

Ernst LF (1978) Drainage of undulating sandy soils with high groundwater tables. J Hydrol 39:1–50

Feddes RA (1987) Crop factors in relation to Makkink reference-crop evapotranspiration. In: Hooghart JC (ed) Evaporation and weather. Proceedings and Information No. 39, TNO committee on hydrological research, The Hague, pp 33–45

Mioduszewski W (2004) Expertise on opportunity change of technology of peat excavation in the Rabinowka peat mine (Michalowo community) (not printed)

Querner EP (1997) Description and application of the combined surface and groundwater flow model MOGROW. J Hydrol 192:158–188

Van Walsum PEV, Veldhuizen AA, van Bakel PJT, van der Bolt FJE, Dik PE, Groenendijk P, Querner EP, Smit MFR (2004) SIMGRO 5.0.1, theory and model implementation. Alterra, Wageningen Alterra Report 913.1

Groundwater Modelling and Hydrological System Analysis of Wetlands in the Middle Biebrza Basin

Mateusz Grygoruk, Okke Batelaan, Tomasz Okruszko,
Dorota Mirosław-Świątek, Jarosław Chormański
and Marek Rycharski

Abstract In the presented approach, a three dimensional finite-difference steady-state groundwater model was applied to analyze the groundwater flow system of the Middle Biebrza Basin. Study contains analysis of hydrogeological and morphological outline of the area, as well as the description of developed groundwater model including conceptual model description, model calibration and sensitivity analysis of parameters. Analysis of volumetric water budget of the model within assumed boundary conditions indicated that groundwater resources of the analyzed part of the Middle Biebrza Basin in approximately 80% come from lateral inflow from the adjacent plateaus. Analysis of spatial distribution of groundwater discharge indicated that the most intensive groundwater inflow to the top peat layer is concentrated within the "Czerwone Bagno", where the peatlands are not degraded and wetland habitats develop naturally, not being directly impacted by drainage ditches and canals.

M. Grygoruk (✉) · T. Okruszko · D. Mirosław-Świątek · J. Chormański
Department of Hydraulic Engineering, Warsaw University of Life Sciences,
ul. Nowoursynowska 159, 02-787 Warsaw, Poland
e-mail: m.grygoruk@levis.sggw.pl

M. Grygoruk · O. Batelaan
Department of Hydrology and Hydraulic Engineering,
Vrije Universiteit Brussel, Brussels, Belgium

O. Batelaan
Department of Earth and Environmental Sciences,
K.U. Leuven, Leuven, Belgium

M. Rycharski
Institute of Technology and Life Sciences, Falenty, Poland

D. Mirosław-Świątek and T. Okruszko (eds.), *Modelling of Hydrological Processes in the Narew Catchment*, Geoplanet: Earth and Planetary Sciences, DOI: 10.1007/978-3-642-19059-9_6, © Springer-Verlag Berlin Heidelberg 2011

1 Introduction

Middle Biebrza Basin (MBB), as part of the Biebrza Valley, belongs to one of the most valuable wetlands within Central Europe. High biodiversity of the these wetlands is expressed by presence of unique mire ecosystems that become habitats for rare plant communities and protected animal species. Certain parts of the MBB (Fig. 1), such as the "Czerwone Bagno", which is one of the core wetland areas of the Biebrza Valley, are strictly protected due to their priceless ecological values and a near-natural hydrological conditions that influence a natural development of wetlands. In this regard, the Valley of Biebrza has been proposed as a reference in the international wetland research (Wassen et al. 2002). However, in nineteenth and twentieth century, human activities have caused alteration of MBB's wetland ecosystems. The hydrographic network of the MBB was changed by Woznawiejski and Augustowski Canals, which were constructed in 1820–1860 (Byczkowski and Kiciński 1991; Maleszewski 1861). Also later, in 1950–1980, some drainage works were executed, that strongly influenced groundwater flow within the MBB. As a result, the mineralization of the peat cover in the surroundings of canals and drainage systems occurred (Mioduszewski et al. 1996). Recently, due to many environmental-restoration strategies and EU projects aimed at reclamation of degraded peatlands, the need of a proper description of water circulation within the MBB occurred. Analysis of mire ecosystem function must involve therefore an appropriate description of groundwater systems (Okruszko et al. 1996; Pajnowska et al. 1984; Wassen 1990). In such analyzes, special regard should be paid to spatial distribution and quantification of groundwater discharge, which is the main factor of fen ecosystems development and function (Wassen 1990).

In this regard, the need for a detailed groundwater flow system analysis of the MBB, especially within the "Czerwone Bagno", through the use of modelling tools, was indicated by several authors, whose works have been based on hydrological, hydrogeological and hydrochemical monitoring (Chormański et al. 2009; Pajnowska 1996; Pajnowska et al. 1984; Stelmaszczyk et al. 2009). As fieldwork within the MBB is complicated due to permanent saturation of the top peat layer and lack of beaten tracks and paths, hydrogeological data is rather poor and limited only to surroundings of the core areas of the basin (Falkowski and Złotoszewska-Niedziałek 2008). Hence, during the last decades, studies on hydrogeology and hydrology of wetlands in the MBB were mostly based on statistical analysis of groundwater level observations and available geological data (Falkowski and Złotoszewska-Niedziałek 2008; Kowalewski et al. 2000; Liwski et al. 1983; Pajnowska et al. 1984; Pajnowska and Poźniak 1991; Poźniak et al. 1983; Żurek et al.1984). The main goal of groundwater modelling studies undertaken so far within the Biebrza Valley (Querner et al. 2010) and the Middle Biebrza Basin was to forecast the response of drained wetlands to planned restoration of artificial canals. For this purpose, the SIMGRO model (Ślesicka and Querner 2000), hydrological models (Bleuten and Schermers 1994; Kubrak and Okruszko 2000) and the MODFLOW model (Książyński et al. 1996) were applied. However, the

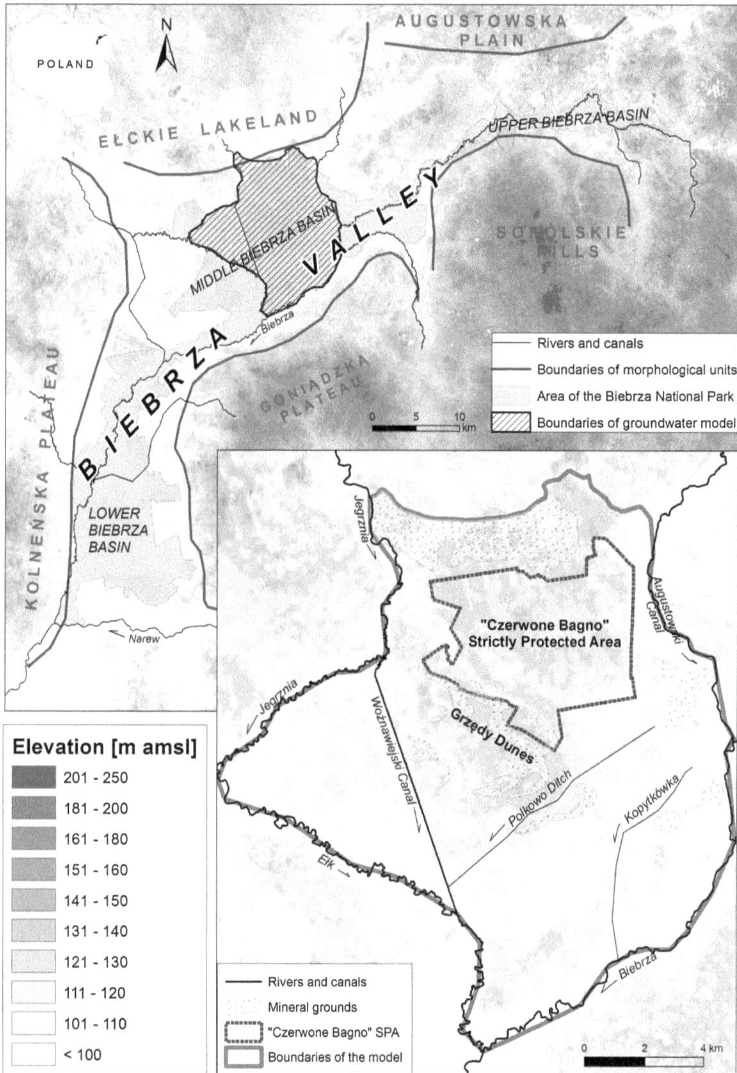

Fig. 1 Elevation and geomorphology elements of the Biebrza Valley, the Middle Biebrza Basin and surrounding areas (following Kondracki 2001 and Banaszuk 2004, modified)

conceptual set-up of those models does not pay significant attention to the groundwater discharge. The limited set of hydrological and hydrogeological data used in setting-up and calibrating these applied models makes those approaches fail to quantify spatially distributed groundwater feeding of valuable peatlands, especially within the "Czerwone Bagno". Such a remark was already given by Żurek (2005). He admitted that groundwater circulation studies undertaken so far within the Middle Biebrza Basin are too general to be used in spatial analysis of groundwater feeding within particular wetlands. Wishing the applied methodology to be comprehensive in mire hydrology assessment, also the vertical upward flow within the mires, especially in the top system, should be analysed (Reeve et al. 2000). Application of the MODFLOW model within the Upper Biebrza headwater catchment in regard to analyse the water feeding of wetlands has been proven to be an appropriate tool (Batelaan and Kuntohadi 2002; van Loon et al. 2009).

The following approach is an attempt of comprehensive description of the groundwater flow system within the Middle Biebrza Basin, herein referred to as the Biebrza Valley extent limited by the Augustowski Canal and Biebrza, Elk and Jegrznia Rivers (Fig. 1). Special regard was given to the analysis of groundwater discharge of near-natural wetlands, located within the "Czerwone Bagno" Strictly Protected Area. Groundwater flow model based on the MODFLOW code (McDonald and Harbaugh 1988) was applied to examine groundwater flow paths and to quantify groundwater seepage. Model calibration included unique, continuous data of groundwater level measurements undertaken in 2007–2010. Those measurements could have been done within the normally inaccessible, strictly protected area, in the framework of project *EEA Grant (PL-0082) "Biodiversity protection of Red Bog (Czerwone Bagno)—relic of raised bogs in Central Europe"*. Main goals of the presented study is: (i) to gather together information on morphology, hydrology and hydrogeology of the MBB as a basis for groundwater modelling; (ii) describe the groundwater flow model set-up and calibration and (iii) analyse the spatial distribution of modelled groundwater levels and groundwater seepage to surficial layer of peat. The conclusions resulting from the presented research become the first step in further analysis, in which understanding of groundwater flow is a crucial aspect of complex and dynamic modelling of the development of mire ecosystems of the MBB. As already recognized in neighbouring parts of the Biebrza Valley (Mioduszewski et al. 2004), results of groundwater modelling become a step forward to the proper water management strategies, which should include complexity of hydrological processes, especially within valuable, natural wetland ecosystems.

2 Study Area

The shallow geology and geomorphology of the Biebrza Valley is the result of complex interactions between accumulation and erosion processes that took place during the last two Pleistocene glaciations and interglacials (Żurek 1984). Due to various intensities of morphogenetic processes within the Pleistocene and Holocene, Biebrza Valley is subdivided into three major units, referred to as "basins": Upper (Northern) Basin, Middle Basin and the Lower (Southern) Basin (Żurek 1984; Fig. 1). Kondracki (2001) includes also a fourth basin, the basin of Wizna, which surrounds the confluence of Biebrza and Narew. The vast depression of the Biebrza Valley is bordered by postglacial plateaus (Kolneńska Plateau (SW), Elk Lake

District (W), Augustowska Plateau (N) and Białostocka (Goniądzka) Plateau (E) (Kondracki 2001). Quaternary sediments within the valley consist of a sequence of sand and till layers. The total thickness of Quaternary sediments reaches there approximately 80 m (Ber 2005; Różycki 1972). The present geomorphology of the Biebrza Valley was heavily impacted by the Wartian deglaciation processes, which led to accumulation of tills (Musiał 1992; Lindner 1988). During the Vistulian frontal deglaciation (approx. 12,000–10,000 years ago), Biebrza Valley became an upper reach of the Biebrza-Narew Ice-Marginal Valley—the main recipient of glacial-fluvial waters in the region, which drove the general outflow from the glacier along the edges of Białostocka Plateau in SW direction (Banaszuk 2001; Galon 1972; Mojski 2005). In the Middle Biebrza Basin, the top-most Quaternary layer consists of sand and gravel deposits that rest—in general—on impermeable strata of tills and silts (Lindner 1988; Mojski 1972; Musiał 1992; Żurek 2005). Higher elevated areas, mainly built up from fine sands and hence referred to as mineral islands (e.g., area of Polkowo and Kopytkowo), are residuals of glacial-fluvial ablation processes (Musiał 1992). However, Żurek (1984) explains those islands as parts of upper accumulation terraces. Since the end of Vistulian periglacial period, small open-water reservoirs temporarily occurred in depressions of sandy underlain areas (Żurek 2005). Residuals of these reservoirs contain gyttia sediments and are currently located at various positions within the MBB. Some authors suspect significant importance of post-glacial vertical movements of sediments for the MBB current relief (Lindner and Marks 1995; Mojski 1972; Musiał 1992; Żurek 1984). Such glacial-tectonic deformations occurred in result of the decreasing pressure of transgressing glaciers (Musiał 1992).

Faults may have occurred especially in strata of till and caused possible vertical movements. Due to this, some sediment layers, even the remains of the recent Vistulian glaciation, may have been deformed. After the Vistulian glaciation, some coarser sands were deposited as dunes due to eolic processes.

Zone of the Grzędy Dunes (see Figs. 1 and 2) cuts the analyzed part of the MBB from north-west towards south-east direction. The dunes are locally elevated more than 10 m above the surface of the peatland. Sandy deposits in the central part of the analyzed area of the MBB, west from the dunes of Grzędy, form a depression, in which the peatlands of "Czerwone Bagno" developed (Figs. 1 and 2). The small slope of the valley, flat relief and the small groundwater slope are the reason, that the area keeps being saturated during all seasons. Mean groundwater levels calculated on the basis of field measurements done in purpose of this study in 2007–2010 within the area of research was almost equal to the ground level. Measured groundwater level magnitudes within the study area in this period varied from 0.5 to 1.1 m. Such conditions have let the organic matter to accumulate during the Holocene. In the MBB the thickness of peat cover is the highest within the "Czerwone Bagno" area and varies up to approximately 4.5 m (Żurek 2005). An overview of the lithology of the area is given in Fig. 2.

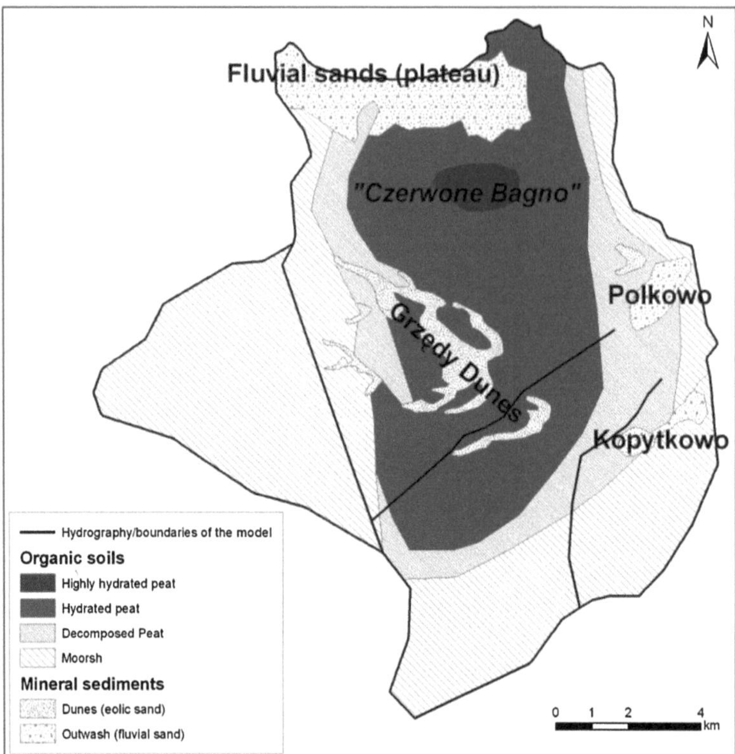

Fig. 2 Geomorphology, soils and lithology of the Middle Biebrza Basin (following Falkowski and Złotoszewska-Niedziałek (2008), Okruszko et al. (1996) and Żurek (1983), modified)

3 Hydrogeology

Outline of hydrogeology of the study area was provided by Falkowski and Złotoszewska-Niedziałek (2008) and Pajnowska (1996). In general, the Valley of Biebrza is the sink of regional surface- and groundwater systems (Pajnowska 1996). Groundwater inflow into the valley occurs in the sandy aquifers from adjacent plateaus. Main directions of groundwater inflow to the Biebrza Valley described as from north towards south, within the valley are impacted by the main rivers and canals. In the neighbourhood of the Biebrza River, along the edges of Białostocka (Goniądzka) Plateau, groundwater flow directions were suspected by Pajnowska (1996) to be more parallel to the course of Biebrza. The most important rivers and canals are in hydraulic contact with the groundwater of the sandy aquifers and drain them (Książyński et al. 1996). The hydrogeological system of aquifers, which strongly determine the groundwater flow and actual hydrological situation of wetlands in the Biebrza Valley, consists of two sandy aquifers: a top

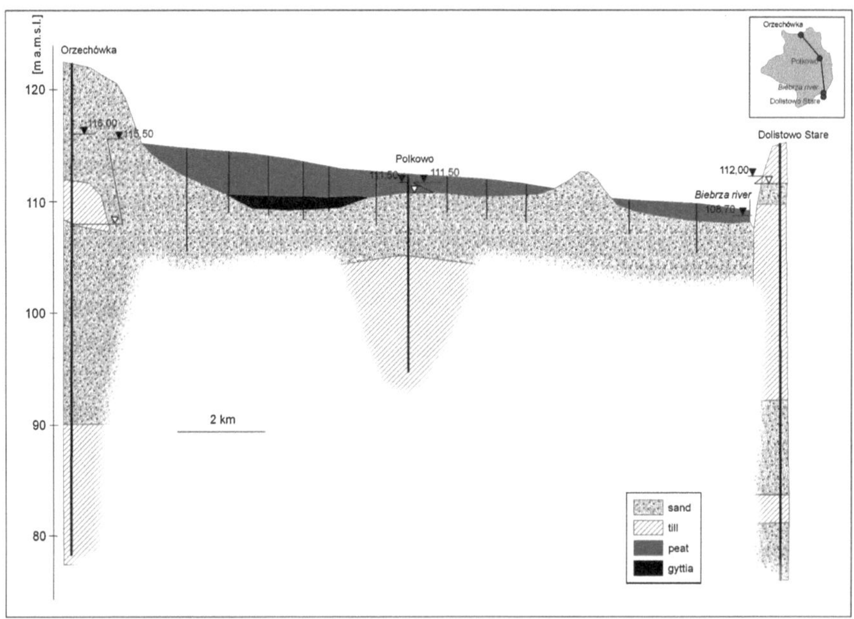

Fig. 3 N–S hydrogeological cross-section through the Middle Biebrza Basin (following Falkowski and Złotoszewska-Niedziałek 2008, extended and modified)

sandy aquifer and a bottom, intra-morainic sandy aquifer (Pajnowska et al. 1984). Top sandy aquifer is semi-confined due to the overlying, low-conductive and constantly saturated organic layer. Within the Biebrza Valley, the top sandy aquifer rests on impermeable strata of till, which forms the separation with the bottom, intra-morainic sandy aquifer (Pajnowska et al. 1984). The intra-morainic aquifer rests on thick strata of impermeable tills and clays of earlier glaciations. However, analysis of piezometric pressures of groundwater within the MBB indicates that there is a good hydraulic contact between both aquifers. As indicated by Pajnowska (1996) and proved by hydrogeological drillings done in purpose of this study, some impermeable strata of till within the MBB were likely eroded during the processes that shaped the ice-marginal Valley of Biebrza after the Vistulian glaciation.

Hence, groundwater of both aquifers is in contact and form one groundwater flow system drained by the Biebrza river. S–E and W–E hydrogeological profiles of the MBB are presented in Figs. 3 and 4. Sandy aquifer within the modelled area is the thickest in the area of Orzechówka and reaches 33 m. The thickness of sandy aquifers decreases to approximately 5 m in between the Grzędy dunes and Polkowo (Fig. 3).

Such a relief of the top sandy aquifer, that decreases from Orzechówka towards Polkowo, with any possibility impacts directions of groundwater flow within this part of study area, making the groundwater flow directions more upward-vertical. In result, the area located south from the "Czerwone Bagno" can be even more

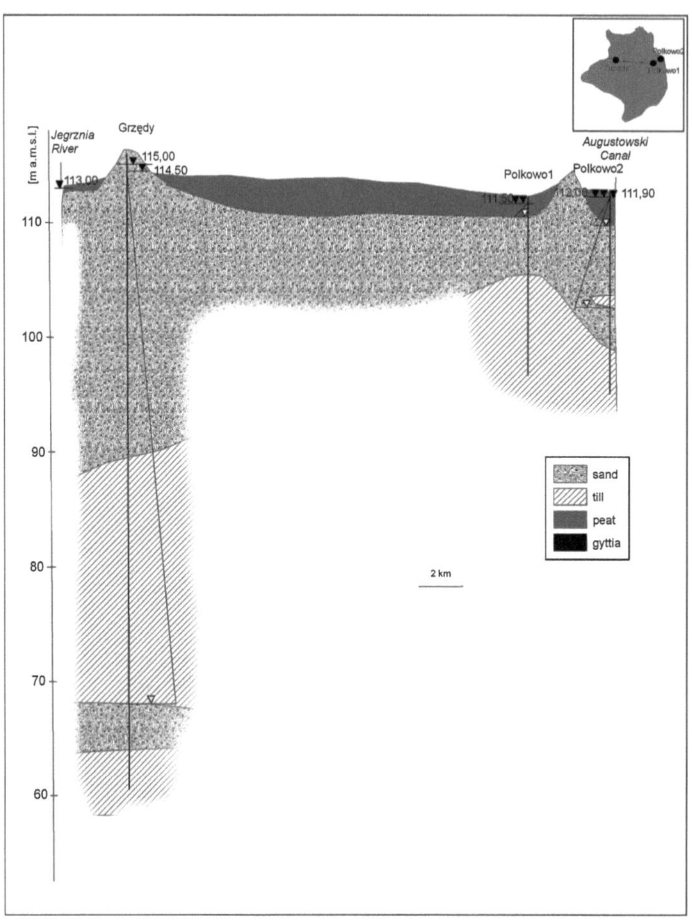

Fig. 4 W–E hydrogeological cross-section through the Middle Biebrza Basin (following Falkowski and Złotoszewska-Niedziałek 2008, extended and modified)

saturated, as groundwater is likely to discharge there more intensively to the overlying, confining organic layer. This layer, that confines the top sandy aquifer within almost whole the study area, consists of various types of peat and gyttia (Żurek 1983). Hydrological processes taking place in this layer are the most important for ecological relations in wetland ecosystems. Due to the low hydraulic conductivity of strata forming this layer, peat of the MBB becomes a confining layer for the underlying aquifers. However, the proper wetland function is strongly dependent of the groundwater feeding of peat from the sandy aquifer. In regard with increased intensity of drainage due to construction of Woznawiejski and Augustowski Canal (modification of the hydrographic network can be found in Byczkowski and Kiciński 1991), peatlands along the main river courses have been decomposed. Moorshing processes have lead to the negative alteration of ecohydrological relations within both—groundwater and surface water fed wetlands within the whole valley of Biebrza (Mioduszewski et al. 1996; Okruszko 1991). Draining impact of canals is suspected to have a broader extent, than only the occurrence of decomposed peaty sediments, although this hypothesis can be verified only on the basis of precise groundwater flow directions and intensity analysis.

4 Groundwater Model

4.1 Model Setup

The presented groundwater model of the MBB is based on the MODFLOW code, versions 2000/2005 (Harbaugh 2005; Harbaugh et al. 2000). The groundwater model domain is divided in grid cells of specified horizontal and vertical resolution.

$$\frac{\partial}{\partial x}\left(K_{xx}\frac{\partial h}{\partial x}\right) + \frac{\partial}{\partial y}\left(K_{yy}\frac{\partial h}{\partial y}\right) + \frac{\partial}{\partial z}\left(K_{zz}\frac{\partial h}{\partial z}\right) + W = S_S\frac{\partial h}{\partial t} \tag{1}$$

Groundwater heads are calculated for nodes, which are the centroids of grid cells. Groundwater heads are calculated by solving Eq. 1, using the finite-difference method (McDonald and Harbaugh 1988), where K_{xx}, K_{yy}, and K_{zz} is the hydraulic conductivity along the x, y, and z coordinate axes, expressed in (L/T); h is the potentiometric groundwater head (L); W is a volumetric flux per unit volume representing sources and/or sinks of water, with $W < 0.0$ for flow out of the groundwater system and $W > 0.0$ for flow in (T^{-1}); S_s is the specific storage of the porous material (L^{-1}); t is time (T) (L—length, T—time). The groundwater model was set up with the GMS 6.0 software, which uses MODFLOW 2000. Further simulations and analysis were done, however, in the USGS's freeware *ModelMuse*[1] (Winston 2009) user interface, using MODFLOW 2005. The horizontal resolution of the model was set uniform over the domain to 100 m. Maximum number of outer iterations (MXITER) and maximum number of inner iterations (ITER1) were defined as 50. The convergence criterion was set up to 0.01 m.

Modelled area equals 182.5 km². Groundwater flow was simulated with use of the Layer Property Flow Package. To represent particular aquifers and types of strata, the model domain was divided vertically into 4 layers (Fig. 5). Elevation of particular layers as well as the relief were obtained from the digitalization of elevation data from topographic maps 1:10000 and 1:25000 and supplied with geodetic measurements (levelling instrument + GPS RTK). The vertical distribution of layers was based on soil and hydrogeological research (Okruszko et al. 1996;

---

[1] http://water.usgs.gov/software/

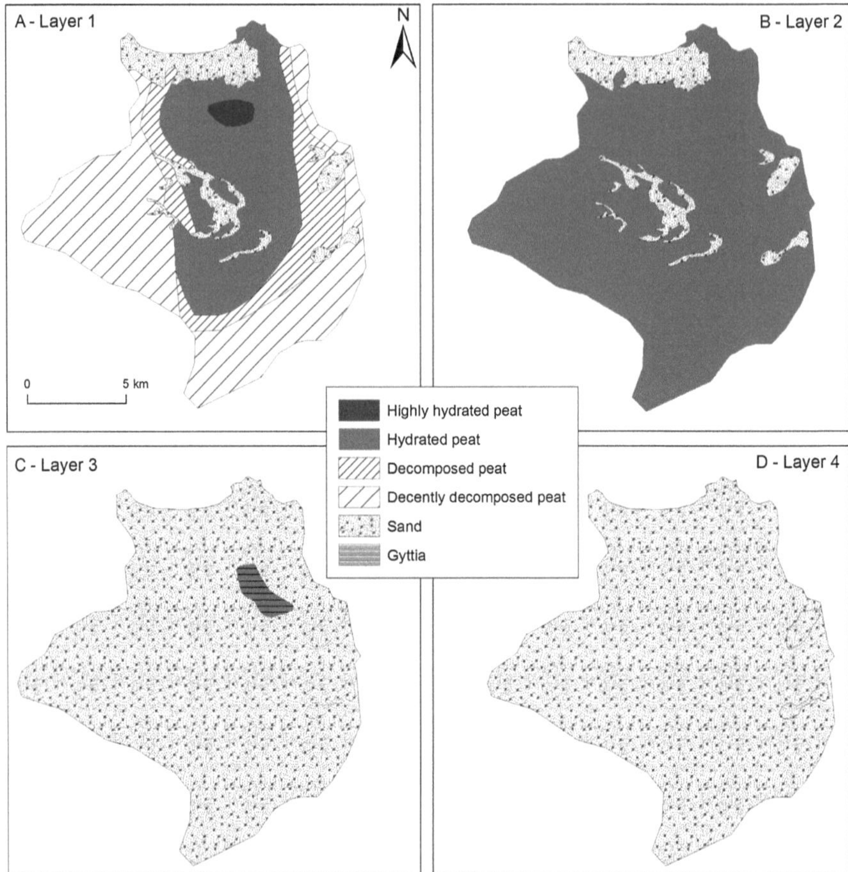

Fig. 5 Spatial distribution of lithology in four MODFLOW grid layers

Pajnowska 1996; Banaszuk and Szuniewicz 2001; Falkowski and Złotoszewska-Niedziałek 2008). The top layer (Fig. 5a) represents the spatial variability of various types of peat and sands. Within the peat extent, different types of organic matter due to the state of its decomposition were taken into account for the first layer characterisation. The thickness of this layer was set up uniform along the model domain and equalled 1 m. The second layer (Fig. 5b) represents hydrated peat that underlay the top organic matter. The total thickness of the first and second layer equals the field examined thickness of organic layer, which varies up to approximately 4.5 m in the core area of "Czerwone Bagno". In areas situated outside of the occurrence of organic matter, the second layer represents sands. The third (Fig. 5c) layer represents various types of sands and the gyttia. Gyttia sediments, that underlay peat, are present in the central part of the modelled area. The fourth layer (Fig. 5d) represents homogenous, as assumed, glacial-fluvial sands.

The First Type Boundary Conditions (CHD MODFLOW Package) were applied to the model along the main rivers and canals. As the presented model was set up in steady-state, the elevation of water applied in CHD cells equalled to the mean groundwater and surface water stages. Along the northern boundary condition, mean values of groundwater elevation measured in the northernmost piezometers were interpolated and applied. In the eastern edge of the northern boundary, the applied CHD value was the mean water level of the Tajno Lake recorded from topographic map.

Along the major rivers and canals the elevation of surface water courses was taken from surface water level measurements (water gauges in Woźnawieś, Kuligi, Ciszewo, Białobrzegi, Dębowo and Osowiec) in 2007–2010. Longitudinal distribution of water elevation in CHD cells was therefore interpolated along the water bodies on the basis of surface water slopes. Some uncertainties as to the hydraulic slope distribution of rivers Jegrznia and Ełk have had to be initially assumed, as achievable data and single measurements might not become typical for the steady state conditions. Distribution of boundary conditions is presented on Fig. 6. Groundwater recharge was specified with the Recharge Package (RCH). As indicated by Pajnowska (1996), the balance of precipitation and evapotranspiration within the MBB in a long time horizon is negative. Facing this fact it is likely that the precipitation water becomes a neglectable source of groundwater recharge in the MBB. On the other hand, evapotranspiration can play significant role for the water balance of surficial lithological layers, impacting phreatic groundwater level. Also due to the low conductivity of peat, after the rainfall phenomena, some water feeds the top system of peatland. Therefore, in the areas of organic soils, the applied value of the groundwater recharge equalled symbolic value of 0.01 mm/day. Areas of sandy dunes and plateaus, covered mostly with pine forests and crop fields were indicated by Falkowski and Złotoszewska-Niedziałek (2008) as important areas of infiltration. Therefore, groundwater recharge value within those areas was parameterized as 0.265 mm/day, as the difference between mean annual precipitation (583 mm) calculated for the whole Biebrza Valley by Kossowska-Cezak (1984) and the mean evapotranspiration (486 mm) calculated by Brandyk et al. (1996) on the basis of meteorological data recorded at the Biebrza station. Due to the fact that the given value of evapotranspiration was calculated not directly for wetlands present within the MBB, groundwater recharge calculated this way is uncertain. Nevertheless, as more accurate data was unavailable, those values were applied. To quantify spatially distributed groundwater discharge, the MODFLOW Drain Package (DRN) was applied to the whole domain of the model, as described by Anderson and Woessner (1999) and Batelaan et al. (2003). Elevation of applied drain bottom was equal to the ground elevation. Conductance of this drain was set up to be very high to avoid unrealistic water table elevations within the groundwater discharge areas (Batelaan et al. 2003). Hydrological processes simulated this way can be found as a surface runoff either the potential availability of water that is consumed by phreatophytic plants. The Drain Package was also applied to simulate draining impact of Kopytkówka River and the Polkowo Ditch (see Fig. 1).

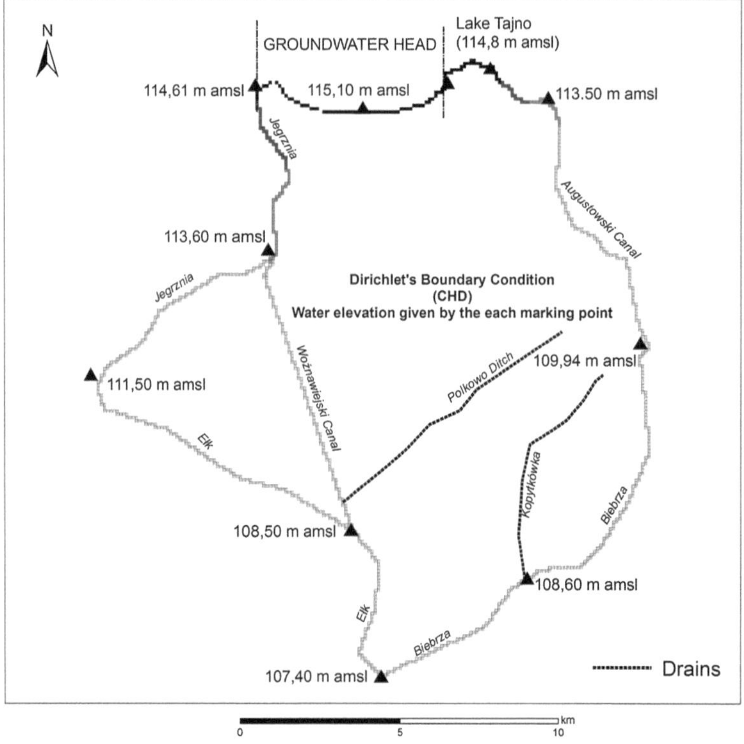

Fig. 6 Boundary conditions of applied model—the MODFLOW CHD package and the MODFLOW DRN package

4.2 Calibration and Sensitivity Analysis

In the first step a basic trial-and-error calibration was used. To adjust parameters of horizontal hydraulic conductivity (K_x and K_y) of particular layers, the PEST algorithm (Doherty 1994) based on inverse modelling technique was applied. K_x and K_y were assumed to be equal for each layer, and the vertical hydraulic conductivity (K_z) was set as $0.1 \times K_x$. Initial parameters of K_x used in PEST calibration were set up as given in references (Bleuten and Schermers 1994; Falkowski and Złotoszewska-Niedziałek 2008; Pajnowska 1996; Ślesicka and Querner 2000; van Loon et al. 2009). Conductivity of the top layer of the model was distributed due to information on the hydrogenic soils in the MBB provided by Okruszko et al. (1996). Conductance of the lowermost sandy layer was set up to be higher than given Falkowski and Złotoszewska-Niedziałek (2008) to provide a better distribution of the lateral inflow to the model domain, which is suspected through the northern boundary of the model. Initial and calibrated parameters are presented in Table 1. Results of the steady-state simulation after calibration of the model were compared to mean annual groundwater heads calculated on the basis

Table 1 Initial and calibrated parameters of hydraulic conductivity used in the PEST inverse modelling estimation

MODFLOW layer	Parameter	Unit	Initial parameter value	Calibrated parameter values
1	K Sands	$m \times s^{-1}$	1.7×10^{-4}	1.39×10^{-4}–2.31×10^{-4}
	K Peat		5×10^{-6}	2.92×10^{-6}–1.15×10^{-5}
	K Moorsh		1×10^{-4}	2.3×10^{-4}
2	K Sands		1.7×10^{-4}	1.39×10^{-4}–2.31×10^{-4}
	K Peat		5×10^{-6}	2.92×10^{-6}
3	K Sands		1.7×10^{-4}	1.15×10^{-3}
	K Gyttia		1.2×10^{-7}	1.2×10^{-7}
4	K Sands		1.7×10^{-4}	1.15×10^{-3}

of field measurements in 33 shallow piezometers within the modelled area during 2007–2010 (Fig. 7).[2] Sixteen piezometers were equipped with automatic Diver® sensors, which recorded data of groundwater elevation every 6 h. In 17 other piezometers groundwater level is measured manually in 10-days interval.

After calibration the root mean squared error (RMSE) of computed groundwater heads versus observations equals 0.26 m. However, the accuracy of the model is much higher, if only piezometers with divers were taken into account in comparison (RMSE = 0.14 m). The explanation is that the manual groundwater level measurements in piezometers not equipped with Divers® are being done only within the period from April up to October, when access to the wetland is possible, in 10-days interval.

In autumn, winter and early spring access to the wetland is limited due to the high water level. Hence, mean values of groundwater head calculated on the basis of manually measured groundwater levels, are representative only for relatively dry periods of the year. It explains the fact that simulated groundwater heads for piezometers without divers are overestimated in general. However, due to the general lack of precise data of groundwater level dynamics within the study area, all the available data were applied in the calibration process, with some constrains and the consciousness of their uncertainty.

The sensitivity analysis was done within the GMS 6.0 application using PEST procedure, set up in the Sensitivity Analysis mode. Composite scaled sensitivities of hydraulic conductivity parameters were calculated (Fig. 8). As shown, results of the analysis with developed model are the most sensitive to the unit change of hydraulic conductivity in the fourth, lowest layer (sand). This high sensitivity is a consequence of the large spatial extent and the extensive thickness of the sandy layer.

The developed model is not sensitive to changes of parameters that express the hydraulic conductivity of gyttias, which can, however, play locally an important role in groundwater flow by affecting inflow to the top peat layer. Also the

---

[2] Data of groundwater heads measured in piezometers without divers come from groundwater monitoring network of the Biebrza National Park.

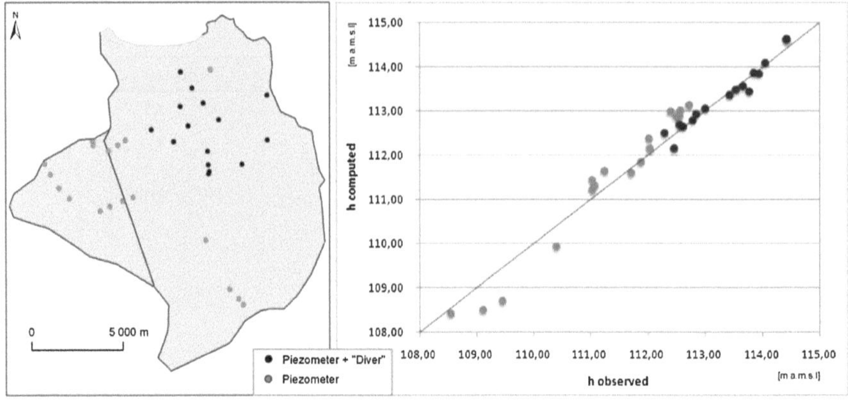

Fig. 7 Comparison of observed versus simulated groundwater heads

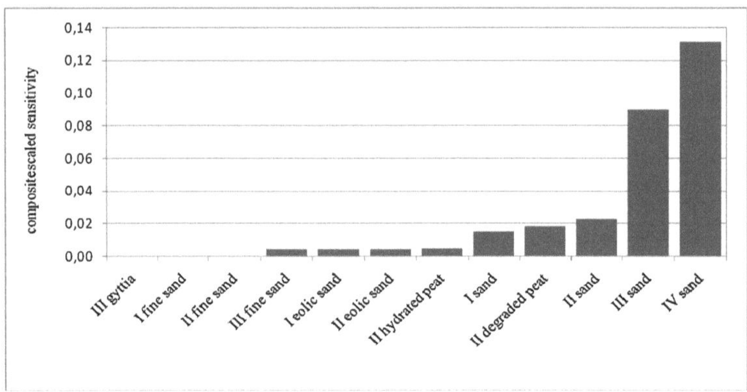

Fig. 8 Composite scaled sensitivity of selected parameters; (description on x axis: layer—stratum)

potential modification of hydraulic conductivity of fine sands deposed on the higher accumulation plateaus of Polkowo and Kopytkowo does not impact decently the results of groundwater head and groundwater flow computation.

5 Results and Discussion

The volumetric discrepancy of the whole model presents neglectable error of 0.01%, which is satisfying. Total inflow of groundwater to the modelled area was calculated as 29,222 m^3/day. Approximately 83% of groundwater volume incoming to the modelled groundwater system (approx. 24,291 m^3/day) come from lateral inflow from adjacent plateaus and infiltration to the modelled area through the constant-head boundaries (northern boundary of the model between

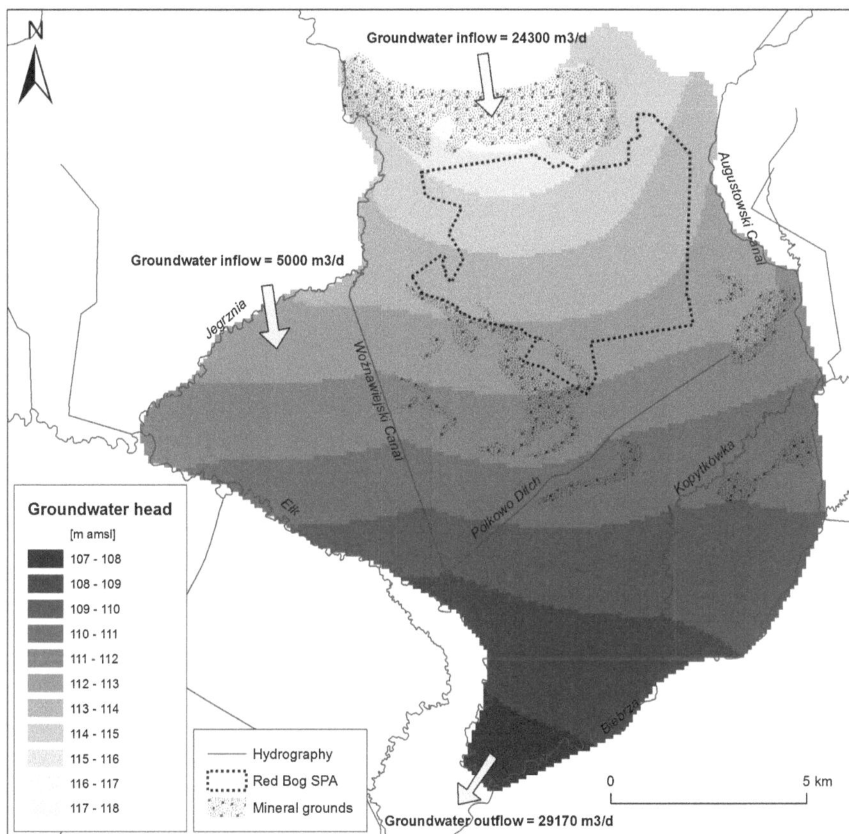

Fig. 9 Modelled groundwater heads

the Jegrznia River and Tajno Lake). Some inflow to the model domain was also observed along the Jegrznia River course, downstream of the Woźnawiejski Canal's outflow. Modelled value of inflow is highly comparable to the one calculated by Pajnowska (1996) which equals 21,874 m^3/day. This fact proves the developed model to provide a truthful values of volumetric water budget components of modelled area. Approximately 17% of the water income to the volumetric budget of whole model comes from the precipitation recharge. Outflow from the model is driven mainly by the constant head cells that represent major streams and equals approximately 29,170 m^3/day. Through applied drains approximately 50 m^3/day of water outflows.

Groundwater heads simulated with the developed model are presented in Fig. 9. In general, the shape and distribution of modelled groundwater head contours is similar to the ones provided in undertaken studies on base of field measurements (Kowalewski et al. 2000 and Pajnowska 1996) and the groundwater modelling (Książyński et al. 1996). Groundwater heads computed for the each of 4 layers are almost equal. Such a state indicates that in general groundwaters of all modelled hydrogeological layers keep in hydraulic contact. This confirms the hypothesis of Pajnowska (1996), which suspected the stretch of the Biebrza Valley within the MBB to be the sink for groundwaters of both shallow and intra-morainic sandy aquifer. Even the gyttia plots (see Fig. 5) does not become a significant obstacle in groundwater exchange between the mineral and organic aquifer. Analysis of groundwater head distribution within the modelled area indicates that the drainage impact of rivers and canals can be observed along almost all water courses. The most intensive draining impact occurs along the Augustowski Canal and Woźnawiejski Canal. Also the modelled draining impact of the river Ełk is quite decent.

However, modelling of groundwater levels within the area of the so-called "Triangle", located between the rivers Jegrznia, Ełk and Woźnawiejski Canal can be strongly affected with uncertainties in the applied boundary condition packages along water courses. Separation of this area from general groundwater flow stream by the draining water courses along all of its bounds makes the groundwater modelling of a saturated zone a complex task there (Książyński et al. 1996). Once the Woźnawiejski Canal was constructed, most of waters of Jegrznia have been captured by the canal and driven throughout the shorter way to River Elk. Therefore, the water elevation of Jegrznia is dependent not only on the hydrological situation of the river up-stream from the canal's outflow but also on the function of the canal. In modelled conditions, Jegrznia River seems to play the role of water supply to adjacent wetlands. Draining impact can also be observed along the Kopytkówka River and the Polkowo Ditch, although due to lacks of data on elevation of the drain's bottom, the low slope of thoses watercourses and regular overgrow with vegetation, analyzed drains seem not to play the significant drainage role in the broad extent.

Modelled elevation of the groundwater table within the areas, where the top layer of the model was defined as mineral soil, is significantly higher. The dunes of Grzędy, as suspected by Falkowski and Złotoszewska-Niedziałek (2008), play an important role in groundwater recharge processes through the infiltration of rainfall. In this area some local water divide appears, which however does not separate the main groundwater flow in the central part of the basin from the flow flux impacted by the Woźnawiejski Canal, west from the Grzędy Dunes.

Through the applied drain package to the whole model domain, groundwater discharge was quantified. In the case presented on Fig. 10, the upward seepage of groundwater from the mineral sandy aquifer to the surface of the peat layer is presented. The most intensive groundwater flux to the surficial layer of the model can be observed in the northern part of peatlands, southwards from the sandy plateaus. Hence, proving the hypothesis of Okruszko (1991), based on the peat type analysis, most of wetlands located within the Czerwone Bagno keep under the strong influence of lateral inflow of groundwater from sandy aquifers. Such a process, which due to construction of Woźnawiejski and Augustowski Canals is likely to be reduced comparing to the one from the before of construction, explains partly the genesis of accumulation of soligenic peat, which is locally more than 4 m thick within the "Czerwone Bagno". The shape of groundwater head contours

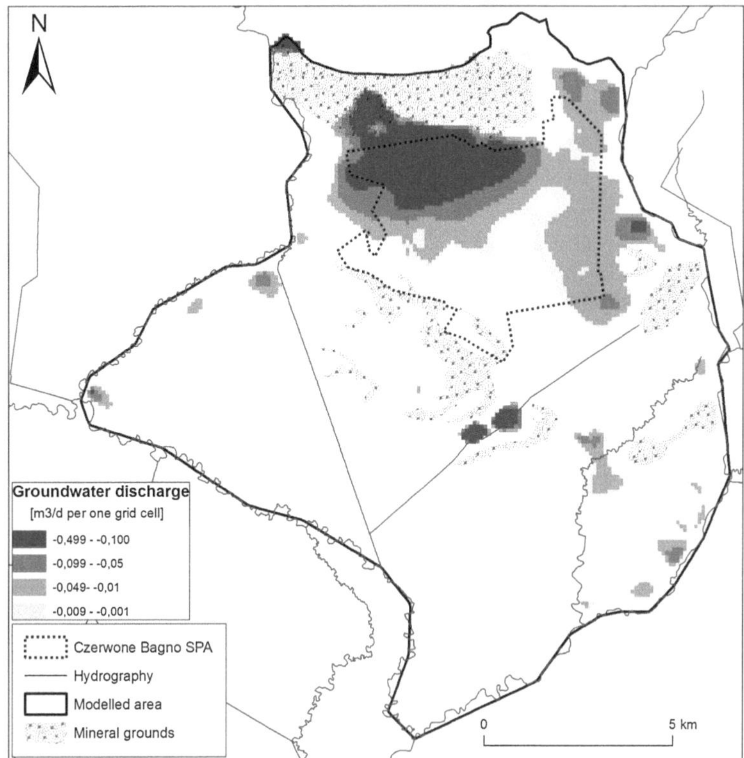

Fig. 10 Spatial distribution of modelled groundwater discharge

within the central part of modelled area indicates that the "Czerwone Bagno" is located on local water-divide. Such a fact has been already suspected by Oświt (1991). The water divide is likely to occur due to the low conductivity of the peat layer that have accumulated there since the end of periglacial period. Low conductive mass of peat could have forced the surface water bodies to the sides of the area. Lateral flow through the peat was being continuously reduced, which could have lead the ombrotrophic conditions to occur in the top system of peatlands in the core areas of the basin.

Inflow of groundwater from sandy aquifer was not sufficiently intensive to dissolve accumulated rainfall water, which could have lead to appearance of acidic conditions. Construction of Woźnawiejski and Augustowski Canals which have changed a general groundwater flow in sandy aquifer with any possibility speeded-up the ombrotrophication processes by reduction of groundwater inflow to the top peat layer. The modelled decline of groundwater discharge intensity that appears according to the gradient from northern boundaries of the Strict Protected Area to core of the "Czerwone Bagno" (Fig. 10) may be therefore considered as the reason of ombrotrophication of central parts of this area.

As the groundwater feeding is still intensive within modelled part of the MBB, the genesis of ombrotrophic habitats in this area should be analysed in conditions of various boundary conditions and model parameters application. In light of presented groundwater discharge analysis results, however, the minerotrophic water seepage appears in a much larger extent than previously suspected (Liwski et al. 1983).

6 Conclusions and Follow Up

1. In light of presented results that have been compared to other studies based on field measurements (Kowalewski et al. 2000; Pajnowska 1996; Pajnowska and Poźniak 1991; Pajnowska et al. 1984; Poźniak et al. 1983) and the groundwater modelling (Bleuten and Schermers 1994; Książyński et al. 1996; Ślesicka and Querner 2000), presented MODFLOW model becomes the useful and comprehensive tool in analysis of groundwater feeding of wetlands.
2. Groundwater inflow from adjacent plateaus becomes the main source of water that feeds wetlands in the Middle Biebrza Basin. Discharge of minerotrophic groundwater appears almost throughout the whole extent of the Czerwone Bagno. Wetlands along the edges of northern mineral sandy plateau are being strongly fed with groundwater from sandy aquifers. Intensity of minerotrophic groundwater discharge decreases southwards, although is still decent along the western boundary of the Czerwone Bagno Strict Protected Area. Low intensity of groundwater discharge in the core area of the Czerwone Bagno indicates the possible appearance of ombrotrophication conditions. This hypothesis however needs to be verified according to the wetland habitat contiguity due to trophic requirements of particular habitats.
3. The "Czerwone Bagno" is located on local water divide of Woźnawiejski Canal and Augustowski Canal. Both canals decently altered groundwater flow within sandy aquifer and peat layer. Hence, prospective strategies aimed at the restoration of water conditions within the Middle Biebrza Basin need to be referred to whole hydraulic system of the area as well as to the groundwater inflow processes—a main source of water for MBB's wetlands. In result of analysis of modelled groundwater heads and groundwater discharge, damming of the Woźnawiejski Canal can force more water to flow into the old course of the Jegrznia River. Facing this fact, broader extent of wetlands along the Jegrznia River can be inundated. Reduction of the draining impact of Woźnawiejski Canal can cause increase of groundwater discharge within the "Czerwone Bagno". Such a process can decently reduce the ombrotrophication of the surficial layers of the peatlands. With use of precise hydrological criterions of wetland function (Okruszko 2005) presented MODFLOW model can be used as a tool in quantitative analysis of response of wetlands to hydrological pressures.

4. Presented approach should be followed by precise measurements of electric conductivity of groundwater on various depths of peatlands. Such a knowledge would let to interpret the relations between rainwater accumulation on top of the peatland with groundwater discharge. Once the groundwater discharge modelling results were confirmed by the field EC measurements in broad extent of the MBB, the developed MODFLOW model will be used to verify hypothesis of Czerwone Bagno wetlands genesis and future development.
5. Field research and measurements of surface water elevation in rivers and canals in various hydrological conditions can benefit the presented approach in some new analysis of groundwater–surface water relations combined with wetland ecosystem response and evolution.

Acknowledgments Research was funded by EEA Grant (PL-0082) "Biodiversity protection of Red Bog (Czerwone Bagno)—relic of raised bogs in Central Europe". Project was developed in Department of Hydraulic Engineering and Environmental Restoration, Warsaw University of Life Sciences. Mirosław Budziński (Biebrza National Park), Mateusz Stelmaszczyk (Warsaw University of Life Sciences), Ewelina Kosela and Anna Socha (Interfaculty Studies of Environmental Conservation, WULS) are gratefully acknowledged for an outstanding help in field research. First author was benefitted from a scholarship of the Flemish Government for Polish-Flanders bilateral exchange. Authors wish to acknowledge Dr hab. Urszula Somorowska and Prof. Waldemar Mioduszewski for their useful comments and the review, which made the paper more robust.

References

Anderson M, Woessner M (1999) Applied groundwater modelling. Blackwell, New York

Banaszuk H (2001) In: About the range of Vistulian glaciations in North-Eastern Poland in result of geomorphological and termoluminescent research (in Polish), Przegląd Geograficzny, p 73

Banaszuk H (2004) General description of Biebrza Valley and Biebrza National Park (in Polish). In: Banaszuk H (ed) Biebrza Valley and Biebrza National Park. Wydawnictwo Ekonomia i Środowisko, Białystok

Banaszuk H, Szuniewicz J (2001) Biotopes and soils (in Polish). In: Mioduszewski W, Kajak A, Gotkiewicz J (eds) Conservation strategy of the Biebrza National Park, General Operate (Project), Warsaw

Batelaan O, Kuntohadi T (2002) Development and application of a groundwater model for the Upper Biebrza River Basin. Annals of Warsaw Agricultural University, Land Reclamation 33:57–69

Batelaan O, de Smedt F, Triest L (2003) Regional groundwater discharge: phreatophyte mapping, groundwater modelling and impact analysis of land-use change. J Hydrol 275:86–108

Ber A (2005) Pleistocene of North-Eastern Poland in regard with deeper underlay and neighboring areas (in Polish). Prace PIG, Warszawa, p 170

Bleuten W, Schermers GT (1994) Hydrological modelling of a part of the Middle Biebrza Basin of the Biebrza Valley. In: Wassen MJ, Okruszko H (eds) Towards protection and sustainable use of the Biebrza wetlands: exchange and integration of research results for the benefit of a Polish-Dutch joint research plan. Institute of Land Reclamation and Grassland Farming Commission of the European Communities Brussels, Department of Environmental Studies, Utrecht University, Utrecht

Brandyk T, Szuniewicz J, Szatyłowicz J, Chrzanowski S (1996) Plant water requirements in hydrogenic areas (in Polish). Zeszyty Problemowe Postępów Nauk Rolniczych, p 432

Byczkowski A, Kiciński T (1991) Hydrology and hydrography of the catchment of Biebrza (in Polish). Zeszyty Problemowe Postępów Nauk Rolniczych p 372

Chormański J, Kardel I, Świątek D, Grygoruk M, Okruszko T (2009) Management support system for wetlands protection red bog and lower Biebrza Valley case study. Hydroinform Hydrol Hydrogeol Water Resour. Proceedings of symposium. JS.4 Joint IAHS & IAH Convention, Hyderabad, IAHS Publ, p 331

Doherty J (1994) PEST; Model—independent parameter estimation. User Manual, 5th edn. Watermark Numerical Computing

Falkowski T, Złotoszewska-Niedziałek H (2008) Geological research for hydrogeological model of water feeding of peatlands of "Red Bog" Reserve (in Polish). Biuletyn PIG, p 431

Galon R (1972) Main stages of Polish lowlands relief genesis (in Polish). Geomorfologia Polski 2 PWN, Warszawa

Harbaugh AW (2005) MODFLOW-2005, the U.S. geological survey modular ground-water model—the ground-water flow process, U.S. geological survey techniques and methods, p 6–A16

Harbaugh AW, Banta ER, Hill MC, McDonald MG (2000) MODFLOW-2000. The U.S. geological survey modular ground-water model—user guide to modularization concepts and the ground-water flow processes. United States Geological Survey Government Printing Office, Reston

Kondracki J (2001) Regional geography of Poland (in Polish). PWN, Warszawa

Kossowska-Cezak U (1984) Climate of the Biebrza ice-marginal valley. Polish Ecol Stud 10:3–4

Kowalewski Z, Skąpski J, Bleuten W (2000) Groundwater in the Central Biebrza Basin. In: Mioduszewski W, Wassen MJ (eds) Some aspects of water management in the valley of Biebrza River. IMUZ, Falenty

Książyński K, Kubrak J, Nachlik E (1996) Numerical simulation of the water cycle in the Middle Biebrza Basin. Zeszyty Problemowe Postępów Nauk Rolniczych, p 432

Kubrak J, Okruszko T (2000) Hydraulic model of the surface water system for the Central Biebrza Basin. In: Mioduszewski W, Wassen MJ (eds) Some aspects of water management in the valley of Biebrza River. IMUZ, Falenty

Lindner L (1988) Stratigraphy and extent of Pleistocene continental glaciations in Europe. Acta Geol Pol 38:1–4

Lindner L, Marks L (1995) Outline of paleogeomorphology in the area of Poland during Scandinavian glaciations (in Polish), Przegląd Geologiczny, p 7

Liwski S, Okruszko H, Poźniak R (1983) Relation between types of hydrogenic soils and hydrogeological conditions in upper and middle Biebrza Basin. Zeszyty Problemowe Postępów Nauk Rolniczych, p 255

Maleszewski W (1861) Drainage of wetlands and meadows in the kingdom of Poland (in Polish). Gazeta Rolnicza, p 14

McDonald MG, Harbaugh AW (1988) A modular three-dimensional finite-difference ground-water flow model, U.S. geological survey techniques of water-resources investigations. Book, p 6

Mioduszewski W, Szuniewicz J, Kowalewski Z, Chrzanowski S, Ślesicka A, Borowski J (1996) Water management on peatlands in the middle Biebrza Basin (in Polish). Biblioteka Wiadomości IMUZ, p 90

Mioduszewski W, Querner EJ, Ślesicka A, Zdanowicz A (2004) Basis of water management in the Valley of Lower Biebrza River. J Water Land Dev, p 8

Mojski JE (1972) Podlaska lowland (in Polish). In: Galon R (ed) Geomorfologia Polski. PWN, Warszawa

Mojski JE (2005) Area of Poland in quaternary an outline of morphogenesis (in Polish). PIG, Warszawa

Musiał A (1992) Study on the glacial relief of northern Podlasie (in Polish). Wydawnictwa Uniwersytetu Warszawskiego, Warszawa

Okruszko H (1991) Wetland kinds in Biebrza Valley (in Polish). Zeszyty Problemowe Postępów Nauk Rolniczych, p 372

Okruszko T (2005) Hydrological criteria in wetlands protection (in Polish). Publications of Warsaw University of Life Sciences, Warsaw

Okruszko H, Szuniewicz J, Kamiński J, Chrzanowski S (1996) Environmental features and potential needs of restoration of the Middle Biebrza Basin (in Polish). Zeszyty Problemowe Postępów Nauk Rolniczych, p 432

Oświt J (1991) Structure, genesis and development of peatlands in the Biebrza Valley (in Polish). Zeszyty Problemowe Postępów Nauk Rolniczych, p 372

Pajnowska H (1996) Hydrogeology and the groundwater inflow of the Middle Biebrza Basin (in Polish). Zeszyty Problemowe Postępów Nauk Rolniczych, p 432

Pajnowska H, Poźniak R (1991) Hydrogeology of the Biebrza Valley and surrounding areas (in Polish). Zeszyty Problemowe Postępów Nauk Rolniczych, p 372

Pajnowska H, Poźniak R, Wiencław E (1984) Ground waters of the Biebrza Valley. Polish Ecol Stud 10:3–4

Poźniak R, Pajnowska H, Skibiński J (1983) Hydrogeological conditions of the middle part of the Biebrza Valley (in Polish). Zeszyty Problemowe Postępów Nauk Rolniczych, p 255

Querner EP, Mioduszewski W, Povilaitis A, Ślesicka A (2010) Modelling peatland hydrology: three cases from northern Europe. Polish J Environ Stud 19:1

Reeve AS, Siegel DI, Glaser PH (2000) Simulating vertical flow in large peatlands. J Hydrol 227:207–217

Różycki S (1972) Pleistocene of central Poland in regard with the past of upper Tertiary (in Polish). PWN, Warszawa

Ślesicka A, Querner E (2000) Modelling studies of groundwater dynamics in Central Biebrza Basin. In: Mioduszewski W, Wassen M (eds) Some aspects of water management in the valley of Biebrza River. IMUZ, Falenty

Stelmaszczyk M, Chormański J, Grygoruk M, Kardel I, Okruszko T, Bartoszuk H (2009) Groundwater chemistry variation in wetland vegetation habitats of the "Red Bog Strict Protected Area". In: Łachacz A (ed) Wetlands—their functions and protection. University of Warmia and Mazury, Olsztyn

Van Loon AH, Schot PP, Griffionen J, Bierkens MFP, Batelaan O, Wassen M (2009) Throughflow as a determining factor for habitat contiguity in a near-natural fen. J Hydrol 379:30–40

Wassen M (1990) Water flow as a major landscape ecological factor in fen development. PhD Thesis, University of Utrecht

Wassen M, Bleuten W, Bootsma MC (2002) Biebrza as geographical reference. Annals of Warsaw Agriculture University, Land Reclamation, p 33

Winston RB (2009) ModelMuse—a graphical user interface for MODFLOW–2005 and PHAST. Techniques and methods 6–A29 USGS, Reston

Żurek S (1983) Geomorphological outline of the Middle Biebrza Basin (in Polish). Zeszyty Problemowe Postępów Nauk Rolniczych, p 255

Żurek S (1984) Relief, geologic structure and hydrography of the Biebrza Valley. Polish Ecol Stud 10:3–4

Żurek S (2005) Habitat conditions of the "Czerwone Bagno" (in Polish). In: 80 lat ochrony ścisłej Czerwonego Bagna w Dolinie Biebrzy. Biebrzański Park Narodowy

Aspiration-Reservation Decision Support System for Siemianówka Reservoir

Adam Kiczko and Jarosław J. Napiórkowski

Abstract This paper presents a Multiple Criteria Decision Support System (DSS), based on the Aspiration Reservation method, for the optimal management of the Siemianówka reservoir. The reservoir is localised on the Narew River upstream the Narew National Park (NNP). The river system under consideration consists of a storage reservoir and a 100 km long River Narew reach, at which end the NNP is located. The goal of the study is to provide decision makers with a tool that would allow the environmental requirements of NNP to be included within the reservoir management policy. The proposed DSS allows to find a trade-off between different reservoir goals, such as: agriculture, fisheries, energy production and wetland ecosystems. The control problem is solved using Receding Horizon Optimal Control Technique.

1 Introduction

The concept of the sustainable development requires that all human activities should be carried out in a balance with surrounding environment. This is especially noticeable in the case of water management where mitigation of negative changes in natural ecosystems has become recently a key issue. In this paper we deal with a problem of reservoir management, having an impact on precious wetland ecosystems. The main difficulty comes from the fact that requirements of ecosystems and economical goals of the reservoir are always in a conflict.

A. Kiczko (✉) · J. J. Napiórkowski
Institute of Geophysics, Polish Academy of Sciences, ul. Księcia Janusza 64,
01-452 Warszawa, Poland
e-mail: akiczko@igf.edu.pl

D. Mirosław-Świątek and T. Okruszko (eds.), *Modelling of Hydrological*
Processes in the Narew Catchment, Geoplanet: Earth and Planetary Sciences,
DOI: 10.1007/978-3-642-19059-9_7, © Springer-Verlag Berlin Heidelberg 2011

This leads to the problem of multi-purpose reservoir control, where the task is to find a trade-off between costs or benefits of certain users. However, at this point the question arises: how to compare different subsystems, having different criteria of performance? For example, the benefits from energy production can be directly assessed in economic terms, while applying such straightforward measures on ecological or social requirements is rather a problematic issue (Labadie 2004).

We present here the control system for case study reservoir Siemianówka. Currently management of this reservoir is performed with use of fixed decision rules, which do not allow easily to adjust the release pattern to present requirements. Our solution is based on Receding Horizon Optimal Control (RHOC), which utilizes optimization techniques, allowing to balance supply for different users. In addition the RHOC is much more flexible than fixed decision rules, as it easily makes use of an actual information concerning river basin state and possible inflow forecasts.

The key issue is the formulation of multi-criteria optimization problem. Therefore, in our study, we consider economic as well ecological requirements, having different performance measures, it is impossible to combine them into a single criterion, like, i.e., cost effectiveness. The simplest and the most common solution to this problem is to use the weighted sum union function. However, e.g., Makowski et al. (1995) showed that such approach almost always is conditioned on the criterion with the highest gradient. Therefore many different solutions to this problem were proposed. The most successful ones were based on such techniques like *Goal Programming* i.e., Can and Houck (1984), Goulter and Castensson (1988), Gandolfi and Salewicz (1991), Yang et al. (1992, 1993) where optimization was constrained to a desirable value for each criterion. This allowed to obtain required trade-off between different criteria. An extension of this concept was proposed in the form of interactive decision support systems, allowing the user to choose any appropriative solution form *pareto*-optimal surface. Such applications were presented by Agrell et al. (1995), Berkemer et al. (1993), Makowski et al. (1995).

For this study we applied the Aspiration Reservation Based Decision Support (ARBDS) method, developed by Wierzbicki (1980) and extended by Lewandowski and Wierzbicki (1989). This is an interactive system, which can provide decision makers with a set of optimal solutions, representing their preferences concerning reservoir goals. It does not give an answer concerning the release, as it is in the case of decision rules, but it allows to investigate the results of the decision, including the impact on natural ecosystems.

2 The Upper Narew Case Study

The Siemianówka reservoir is localised on the Narew River in North-East Poland. Downstream to a rich wetland ecosystem, enclosed within Narew National Park (NNP), is situated. The NNPs flora consists of more than 600 species of vascular

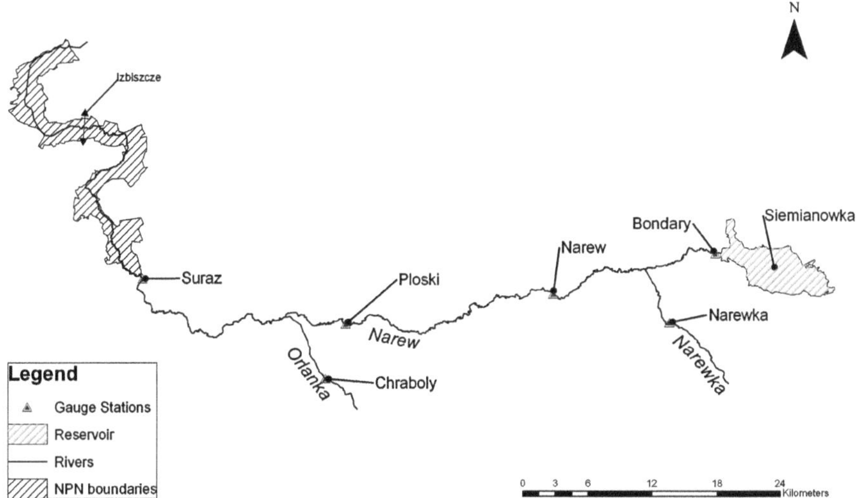

Fig. 1 Schematic map of the study area

plants, including many protected varieties. Park wetland areas provide habitats for about 200 bird species, being one of the most important stop-over points for migrating birds. Due to its unique features, the NNP is an important site in the European Network of Natura 2000 (Dembek and Danielewska 1996).

Freshets which in most of the other regions might cause significant threat, in NNP are the part of a natural hydrological cycle. It can be seen that the localisation of towns and villages follows the inundation zone border. However, in recent years, alarming changes have been observed in the Upper Narew River hydrological regime, manifesting in a reduction of mean flows and shorter flooding periods. This results in a serious threat to rich wetland ecosystems. Local climate changes are one of causes of those changes. Mild winters combined with a reduction in annual rain levels have resulted in a reduction of the valleys groundwater resources. However, recent human activities also have had a significant influence on the deterioration of wetlands water conditions. River regulation work performed in the lower river reach has lowered water levels. Additionally, the Siemianówka reservoir has had also an important impact on water conditions, causing a reduction in flood peaks (Fig. 1).

Adaptation of Siemianówka reservoir management to improve water conditions at protected areas seems to be necessary, especially from the beginning it has a noticeable impact on these ecosystems. Okruszko et al. (1996) suggested that NNP should become the main reservoir recipient. Requirements of other reservoir users cannot be also neglected and in addition to fulfilling wetlands demands the following management goals have to be also included:

- Irrigation of grasslands for agriculture,
- Flood protection of adjacent to the reservoir areas, up to Narewka tributary,

- Fishery at reservoir,
- Energy production from water turbines.

The goals are introduced in form of desirable trajectories for measures describing fulfilling of this goals. In case of wetland ecosystem, which according to Junk et al. (1989), Tockner et al. (2000) depends largely on river flow conditions, and particularly, on flooding, it is assumed that water levels at the inflow are the main factor that determines general conditions. On the base of an experience of Banaszuk et al. (2002), Okruszko and Kiczko (2008) it was possible to estimate its favourable pattern: $\hat{H}(t)$. In the control system the criterion for wetland's requirements was the value of water level below this required one.

Flood protection is seen here as actions aimed at minimization of potential loses. As for this region only arable land could be affected, it is assumed that all costs related to the flooding are proportional to an inundation area. Therefore flood loses criterion is conditioned on admissible flooding area $\hat{A}$ being a nonlinear function of the reservoir's release.

Irrigation objectives of the Siemianówka reservoirs apply only to downstream areas, localised in the its closest neighbourhood. Requirements of agriculture, concerning reservoir operation, are reduced here to desirable reservoir's outflow: $\hat{U}_A$, determined by the current reservoir's management instruction (BIPROMEL 1999) and the deviation from this value is used here as the criterion. In case of fisheries possibly high storage is favourable during spawning period and smaller for catching. Having in mind reservoir capacity limits "optimal" value $\hat{S}(t)$ was shaped according to suggestions from the reservoir management instruction. As the reservoir storage criterion a value of deviation from $\hat{S}(t)$ is used. A performance of hydro-power plant depends directly on the reservoir's release and in the control problem criterion was related to the turbines capacity $\hat{U}_E$. Here the criterion is a value of discharge being below the $\hat{U}_E$.

More details concerning evaluation of these goal trajectories for the Siemianówka reservoir might be found in papers: Kiczko et al. (2008) and Kiczko (2008).

To analyse the influence of reservoir on the river system and dependent subsystems, like wetlands, it was necessary to develop an integrated model for Narew River. Because of relatively small size of the reservoir, compared to the used time scale, its dynamics is described using a simple balance equation:

$$\frac{dS(t)}{dt} = S(t) - U(t) + R(t) - P(t) \tag{1}$$

where: $S(t)$—reservoir storage (m^3) at the time t (s), $U(t)$—release amount (m^3/s) to the river (control variable), $R(t)$—inflow to the reservoir (m^3/s), $P(t)$—evaporation from the reservoir surface (m^3/s). In addition reservoir storage $S(t)$ should not exceed minimal storage value $S_{\min}$ and maximal $S_{\max}$ for all t periods. Similar constraints are imposed on $U(t)$, which has to be higher than $U_{\min}$ and smaller than $U_{\max}$:

$$S_{\min} \le S(t) \le S_{\max}$$
$$U_{\min} \le U(t) \le U_{\max} \tag{2}$$

The water mass between the reservoir and protected areas of NNP is transferred through a nearly 100 km long river reach. Therefore inflow to the NNP—Q (m³/s) is a result of reservoir outflow U transformation along this reach and is also shaped by lateral inflow, denoted as L (m³/s), which in the case of high flows has more than 50% share in the total mass. Unfortunately flow records for lateral inflow are available only for one tributary. As a result the significant part of the water mass comes from uncontrolled sub-basins. This increases a complexity of the reservoir control problem, as modelling of such external sources involves high uncertainty.

A flow routing was modelled using the a Multiple Input Single Output (MISO) Discrete Transfer Function (TF) (see Romanowicz et al. 2007 for more details). This method, although simplified, allows to achieve a computationally efficient description of a river flow process in a lumped form. For the Upper Narew River reach the model has the following structure:

$$Q_k = \frac{B(z^{-1})}{A(z^{-1})} U_k + \frac{C(z^{-1})}{A(z^{-1})} L_k^n \tag{3}$$

where Q_k—water flow (m³/s) at protected and/or agricultural sites, L_k^n lateral inflow to the river system from Narewka tributary (m³), $A(z^{-1})$, $B(z^{-1})$, $C(z^{-1})$ are polynomials of a backshift operator z^{-1}, and k denotes time step. Coefficients of the polynomials were identified using the Captain Toolbox Young et al. (2004). It is important to note that in the presented formulation $\frac{C(z^{-1})}{A(z^{-1})}$ explains not only flow transformation of L^n but also unmeasured components of total lateral inflow, linearly correlated with L^n.

As water conditions at wetland sites are conditioned here on water levels, we use an inverse rating curve relation to transform discharge Q_k into H_k.

3 Inflow Forecasts

The applied methodology requires certain assumptions concerning future values of inflow to the river system R_k and L_k^n. The biggest difficulty in the case of the Upper Narew comes from the fact, due to its lowland character, that most of the river flow is driven through groundwater supply. This causes that interaction between precipitation impulse and outflow is extremely hard to be identified. Therefore, in this study, we have decided to produce inflow forecasts on the basis of a knowledge concerning seasonal pattern in a behaviour of the river system. The simplest way would be to use a daily mean values from historical records as forecast. On such basis current management rules of the Siemianówka reservoir were evaluated. However, such a method does not allow for direct adaptation of the reservoir work

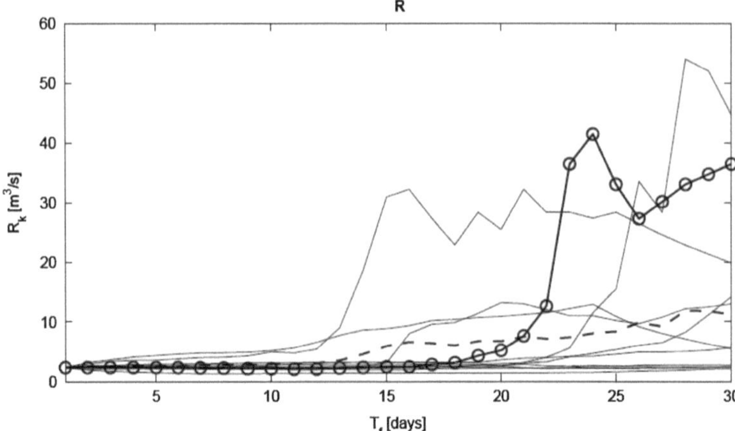

Fig. 2 An example forecast obtained with k-NN method for Bondary; *solid lines* stand for foretasted trajectories, *dashed lines* for an expected ones and *lines with circles* for observations

for changing inflow condition, reducing, in result, performance of the whole control system.

In this paper we perform an inflow forecast by means of the so-called nearest neighbour technique. This is a non-parametric method, introduced into the hydrology by Karlsson and Yakowitz (1987). It has been widely used to forecast inflows from uncontrolled basins (Napiórkowski et al. 1999; Piotrowski et al. 2004).

In the proposed approach, one searches for the K points in Euclidean phase space, representing K trajectories of historical sequences of flows with an "embedding dimension" p, that are the most similar (in the sense of the smallest Euclidean distance) to the point representing the current situation. Then the selected trajectories are applied to flow forecasting for the assumed time horizon. The inflow forecast is calculated as a mean of the trajectories K.

Numerical experiments of the control model showed that this methodology improves the overall objective function value by 30% approximately, compared to the use of daily mean values of discharge as inflow predictions. In Fig. 2 an example forecast for flows at the Bondary river gauge is presented.

4 Optimization Problem

The control problem is formulated according to the Receding Horizon Optimal Control, broadly described by Castelletti et al. (2008). The amount of release— U is computed at each decision step, taking into account the present system state and inflow forecasts. For Siemianówka reservoir the 1 day decision step, recommended in the reservoir management instruction BIPROMEL (1999), is applied.

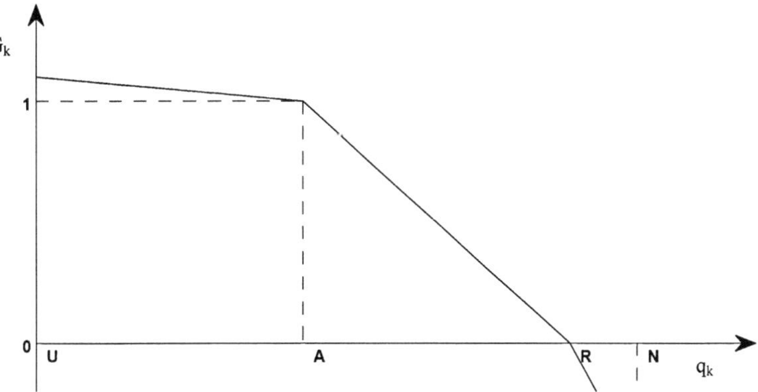

Fig. 3 The Component Achievement Function (CAF); U denotes Utopia point, A Aspiration point, R Reservation point, N Nadir point

Within the RHOC the optimisation problem is solved for a chosen, finite time horizon, reflecting the information concerning future, possible inflows to the system (R and L). Usually it exceeds the duration of a single release disposal, which means that a solution of optimisation problem is evaluated for a whole forecasted period. However, according to this concept only the first disposal is applied as the whole procedure, is repeated at the next time step. The finite time horizon requires to include an additional criterion for a system end-state. In this application it is based on reservoir storage S at the last optimization time step, whose desirable value $\hat{S}_{\mathrm{end}}(t)$ is equal to this for fisheries $\left(\hat{S}(t)\right)$ requirements.

To find an optimal reservoir release at a given time step, that would allow to find a favorable trade-off between all this criteria, the Aspiration Reservation Based Decision Support System (ARBDS) is applied (Lewandowski and Wierzbicki 1989). The control system takes a form of an interactive decision support system, as the reservoir operator takes part in the optimization process.

The ARBDS is based on a special criteria scaling technique, using a Component Achievement Function (CAF). We do not give details of this approach here since a rather comprehensive discussion can be found in Granat and Makowski (2000). In a general form, ARBDS is defined as a maximization of the following function F:

$$F = \min_{1 \leq \kappa \leq n} G_\kappa\left(q_\kappa, \overline{q_\kappa}, \underline{q_\kappa}\right) + \varepsilon \sum_{\kappa=1}^{n} G_\kappa\left(q_\kappa, \overline{q_\kappa}, \underline{q_\kappa}\right) \tag{4}$$

where κ stands for a criterion number, G_κ for the κth CAF, q_κ denotes criteria functions, $\overline{q_\kappa}, \underline{q_\kappa}$ are the aspiration and reservation levels, respectively, and ε— small number ($\sim 10^{-3}$–10^{-5}). The criteria functions q_κ are formulated here as deviation from preferred trajectory for reservoirs users, i.e., $\hat{H}(t)$, $\hat{A}$, $\hat{U}_\mathrm{A}$, $\hat{S}$, $\hat{U}_\mathrm{E}$ and the CAF function has a form presented in Fig. 3.

The Utopia and Nadir points represent respectively the best possible solution for a given criteria and the worst one. During the optimization process the operator adjusts $\overline{q_\kappa}, \underline{q_\kappa}$ levels modifying the shape of CAF, expressing this way preferences concerning the given criterion.

In this study the optimization is performed using the Differential Evolution algorithm, introduced by Storn and Price (1997).

5 Results

The work of Decision Support System for Siemianówka is presented here on a simple management scenario. We analyzed the possible reservoir's operation during one spring freshet event, observed in 1986 year. The optimization was performed for the 1 month long forecast. During the optimization process optimal outflow trajectories for this whole time horizon were computed, however, to reduce the problem dimensionality it was assumed that there are only 4 discharge disposals, where the first one is performed for the 1 day and the rest three for the longer 10 days periods. It is important to note that this simplification affects only the optimization stage and do not have direct impact on a whole control process, as according to the RHOC concept at the next step (1 day step) the whole procedure is repeated, giving the 1 day resolution.

Example results are shown in Fig. 4. Interactive adjustment of reservation and aspiration levels for each criteria allowed to evaluate different solutions, being optimal in a Pareto sense—i.e., such as that it is impossible to improve any of criterion without worsening the rest. The first solution (no. 1) was computed for CAFs marked with solid lines. For the second one, the aspiration and reservation values for the wetland's criterion were reduced. It results in improvement of this criterion, worsening the others. On such basis it is possible to find any desirable trade-off between all criteria.

6 Conclusions

We present the Multi Criteria Decision Support System for Siemianówka reservoir. Our idea was to provide a user with a tool that would allow to find such a reservoir release pattern that keeps in balance economic and ecological demands. It has an interactive form, as we believe that in the case of such different goals as: flood protection, energy production, agriculture, fisheries and protection of natural ecosystems, any rigid solution should not be imposed. At each decision stage our Decision Support system allows to find a trade-off between criteria, reflecting the preferences of the decision maker.

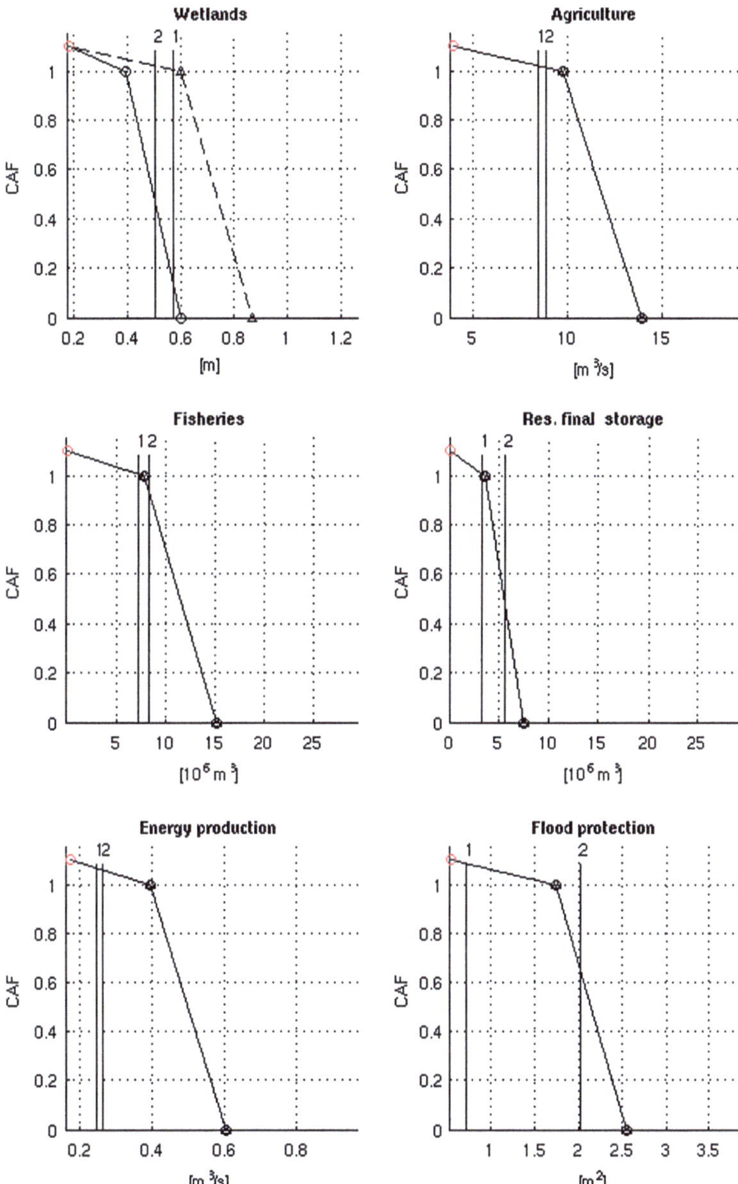

Fig. 4 Multi-criteria analysis of the Siemianówka control problem for 1986 freshet event; *Numbers 1* and *2* stand for a solution number. The CAF function is marked with *dashed line* for a first solution and with a *solid line* for a second one

References

Agrell P, Lence B, Stam A (1995) An interactive multi-criteria decision model for reservoir management the Shellmouth reservoir case. Technical report, International Institute for Applied Systems Analysis, Laxenburg, Austria

Banaszuk P, Banaszuk H, Czubaszek R, Jaros H, Jekatierynczuk-Rudczyk E, Kondratiuk P, Micun K, Roj-Rojewski S, Wysoka-Czubaszek A (2002) Narew National Park protection plan. Technical report, Narew National Park

Berkemer R, Makowski M, Watkins D (1993) A prototype of a decision support system for river basin water quality management in Central and Eastern Europe. Technical report, International Institute for Applied Systems Analysis, Laxenburg, Austria

BIPROMEL (1999) Siemianowka reservoir—water management rules (pol). Technical report, Bipromel, Warszawa

Can E, Houck M (1984) Real time reservoir operations by goal programming. J Water Resour Plan Manag 110(3):297–309

Castelletti A, Pianosi F, Soncini-Sessa R (2008) Water reservoir control under economic, social and environmental constraints. Automatica 44(6):1595–1607

Dembek W, Danielewska A (1996) Habitat diversification of the Upper Narew basin between the Siemianówka reservoir and Suraż (pol.—Zroznicowanie siedliskowe doliny Gornej Narwi od zbiornika Siemianowka do Suraza). Zeszyty Problemowe Postepów Nauk Rolniczych, p 428

Gandolfi C, Salewicz K (1991) Water resources management in the zambesi valley: analysis of the lake kariba operation. In: Van de Ven F, Gutknecht D, Salewicz K (eds) Hydrology for the management of large rivers. IAHS, St. Louis

Goulter I, Castensson R (1988) Multi-objective analysis of boating and fishlife in Lake Sommen. Water Resour Dev 4(3):191–198

Granat J, Makowski M (2000) Interactive specification and analysis of aspiration-based preferences. EJOR 122(2):469–485 available also as IIASA's RR-00-09

Junk W, Bayley P, Sparks R (1989) The flood pulse concept in river-floodplain system. Canian special publication of fisheries and aquatic sciences

Karlsson M, Yakowitz S (1987) Nearest-neighbour methods for nonparametric rainfall-runoff forecasting. Water Resources Research (USA)

Kiczko A (2008) Multi-criteria decision support system for Siemianówka reservoir under uncertainties. Technical Report IR-08-026. International Institute for Applied Systems Analysis, Laxenburg

Kiczko A, Romanowicz R, Napiórkowski J, Piotrowski A (2008) Integration of reservoir management and flow routing model—Upper Narew case study. Publs Inst Geophys Pol Acad Sci E-9(405) pp 41–56

Labadie J (2004) Optimal operation of multireservoir systems: state-of-the-art review. J Water Res Plan Manag, pp 93–111

Lewandowski A, Wierzbicki AP (1989) Aspiration based decision support systems. In: Lecture notes in economics and mathematical systems, no. 331, Berlin

Makowski M, Somlyody L, Watkins D (1995) Multiple criteria analysis for regional water quality management: the Nitra River case. Technical Report WP-95-022, International Institute for Applied Systems Analysis, Laxenburg

Napiórkowski J, Kozlowski A, Terlikowski T (1999) Influence of inflow prediction on performance of water reservoir system. Water industry systems: modelling and optimization applications, p 39

Okruszko T, Kiczko A (2008) Assessment of water requirements of swamp communities: the river Narew case study. Publ Inst Geophys Pol Acad Sci E-9(405):27–39

Okruszko T, Tyszewski S, Pusłowska D (1996) Water management in the Upper Narew River (pol.—Gospodarowanie zasobami wodnymi Górnej Narwi). Zeszyty Prob Post Nauk Rol (428)

Piotrowski A, Rowiński P, Napiórkowski J (2004) River flow forecast by means of selected black box models. In: River Flow 2004

Romanowicz R, Kiczko A, Pappenberger F (2007) A state dependent nonlinear approach to flood forecasting. Publs Inst Geophys Pol Acad Sci E-7(401) pp 223–230

Storn R, Price K (1997) Differential evolution—a simple and efficient heuristic for global optimization over continuous spaces. J Glob Optim 11:341–359

Tockner K, Malard F, Ward J (2000) An extension of the flood pulse concept. Hydrol Process 14:2861–2883

Wierzbicki AP (1980) The use of reference objectives in multiobjective optimisation. In: Fandel G, Gal T (eds) MCDM theory and application, number 177 in lecture notes in economics and mathematical systems, New York

Yang Y, Burn D, Lence B (1992) Development of a framework for the selection of a reservoir operating policy. Can J Civil Eng 19:865–874

Yang Y, Bhatt S, Burn D (1993) Reservoir operating policies considering release change. Civil Eng Syst 10:77–86

Young P, Taylor C, Tych W, Pedregal D, McKenna C (2004) The Captain Toolbox. Centre for Research on Environmental Systems and Statistics, Lancaster University

Hydraulic Conditions of Flood Wave Propagation in the Valley of the Narew River after the Siemianówka Dam Overtopping Failure

Janusz Kubrak, Michał Szydłowski and Dorota Mirosław-Świątek

Abstract In this paper, the results of the dam break wave forecast in the valley of the Narew River, after the failure of the Siemianówka Reservoir earth dam, are presented. The Siemianówka Reservoir was built in 1991 due to the planned agricultural management of the Narew Valley. It is located in the Upper Narew Catchment as a typical lowland reservoir of a low mean depth, equal to 2.5 m and the principal spillway of 145.00 m above sea level. The reservoir is located about 82 km upstream of the border of the Narew National Park and considerably influences the hydrological regime of the Narew River within that park. The total capacity of the reservoir is 79.5 million m^3. The presented example of calculations for the Siemianówka Reservoir underlines the most important issues related with the forecast of the dam break, that is: the determination of maximum water levels and discharges in the valley downstream of the failure dam, the calculation of wave-front occurrence time at particular points of the valley, the inundation range along with the longitudinal profiles of water levels and velocities of water on inundated areas. The knowledge of the water depths on inundated areas, the duration of inundation and maximum water velocities enable as well the estimation of the damage degree on inundated areas. As it results from the calculations, the washout process of the Siemianówka Reservoir earth dam, which is 6 m high, happens very quickly, and after about 2 h the top of the triangular breach reaches the dam bed. The discharge from the reservoir after dam failure drops from 465 to 304 m^3/s and the water level in reservoir decreases 1.36 m as fast as within 300 h. The occurrence of wave in the

J. Kubrak (✉) · D. Mirosław-Świątek
Department of Hydraulic Engineering, Warsaw University of Life Sciences,
Nowoursynowska Str. 159, 02-787 Warszawa, Poland
e-mail: Janusz_Kubrak@sggw.pl

M. Szydłowski
Departament of Hydraulic Engineering, Gdansk University of Technology,
Gdansk, Poland

D. Mirosław-Świątek and T. Okruszko (eds.), *Modelling of Hydrological Processes in the Narew Catchment*, Geoplanet: Earth and Planetary Sciences,
DOI: 10.1007/978-3-642-19059-9_8, © Springer-Verlag Berlin Heidelberg 2011

valley causes the increase of water levels by 0.1–4 m in comparison to the water level before the dam break, and a 4-time increase of discharge rate in relation to the design flow at the cross-section of the dam of the Siemianówka Reservoir. The velocity of the wave-front propagation in the Narew Valley does not exceed 5 km/h.

1 Introduction

Dam failures are not always caused by human errors (De Wrachien and Mambretti 2009). Contemporary engineering craftsmanship has almost completely eliminated the risk of a dam breakdown caused by a construction fault. The greatest danger that exists nowadays, is related to the recognition of geological conditions, however, if all contemporary requirements are fulfilled in this respect, that danger can be considered negligible. A low probability of a dam break does not indicate, however, that it is completely unreal. The damage to the dam causes a sudden release of water stored in the reservoir. The flood wave, which is then created, propagates in the river valley with the velocity and the discharge considerably exceeding the values of those parameters for natural floods (Gallegos et al. 2009; Kubrak 1989; Londe 1981). The flood wave, as a mounding wave, is characterized by a small time delay of wave-peak after the occurrence of wave-front. In practice, the possibilities to mitigate the consequences of wave propagation, created as a result of dam break and an uncontrolled water outflow from reservoir, are limited and generally aimed at improving the conditions of water runoff. This is the reason why already at the phase of dam design the parameters of water flow are forecast as well as the area of inundation. In Poland, the need to forecast the consequences of such disasters is stated by the Order of the Ministry of Environment Protection, Natural and Forest Resources, dated 20 December 1996. This order introduces also the need to create the management plans for the river valley downstream of the dam and the scenarios of rescue and evacuation actions in case of the disaster occurrence.

The forecast of celerity in the valley, caused by the dam failures, can be realized by numerical simulations (Arico et al. 2007; Chanson 2009; Fread 1977, 1984, 1985; Kubrak 1989; Soarez Frazão and Zech 2002; Macchione 2008; Szydłowski 1998, 2003). Physical modeling of a mounding wave propagation in the river valley is not applied in practice on account of the scale of this phenomenon and the variability of conditions initiating the dam break. On the base of the calculation results, the evacuation and warning plans are elaborated for people in the case of potential threat of flooding, caused by the dam break. Basic information derived from the numerical modeling of hydraulic results of dam break includes:

- maximum water level elevations in the valley, downstream of the destroyed dam
- the time of wave-front occurrence at individual points of the river valley
- the area of created inundation and water velocity.

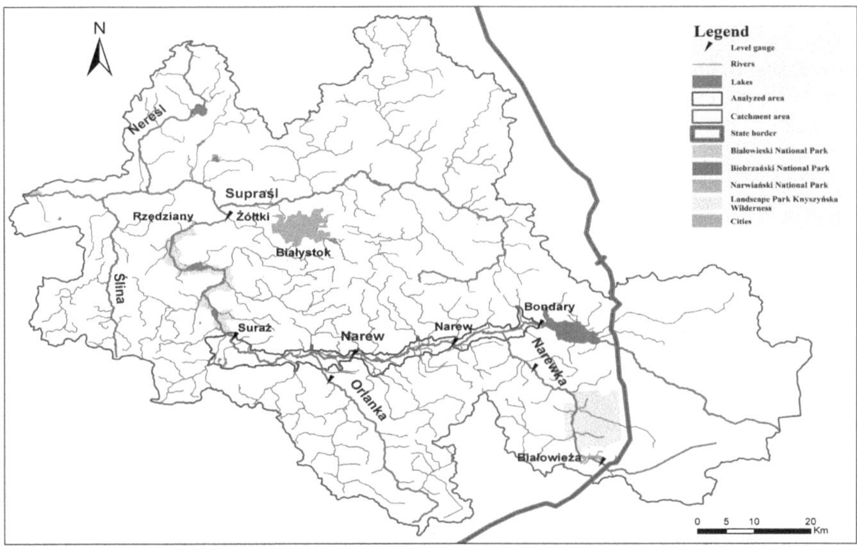

Fig. 1 Stream network of the Narew River

In this paper, the results of the dam break wave forecast in the valley of the Narew River, after the failure of the Siemianówka Reservoir earth dam, are presented.

2 Materials and Methods

2.1 Study Area

The Upper Narew Catchment has a well-developed stream network. Hydrological regime of the Narew River is typical for lowland rivers of Middle Europe. It is characterized by a single winter flood caused by snow cover thawing and by a regular summer runoff. The main left-side tributaries to the Narew River up to the estuary of the Biebrza River are: Narewka, Orlanka and Ślina rivers, while the right-side tributaries include the rivers of: Supraśl and Nereśl (Fig. 1). The Siemianówka Reservoir was built in 1991 due to the planned agricultural management of the Narew Valley. It is located in the Upper Narew Catchment as a typical lowland reservoir of a low mean depth, equal to 2.5 m and the principal spillway of 145.00 m above sea level (Sokolowski 1999). The reservoir is located about 82 km upstream of the border of the Narew National Park and considerably influences the hydrological regime of the Narew River within that park. The total capacity of the reservoir is 79.5 million m^3.

The level of the crest of the reservoir front dam was designed to be 147.00 m a.s.l., that is 2 m above the normal water level, which is equal to 145.00 m a.s.l..

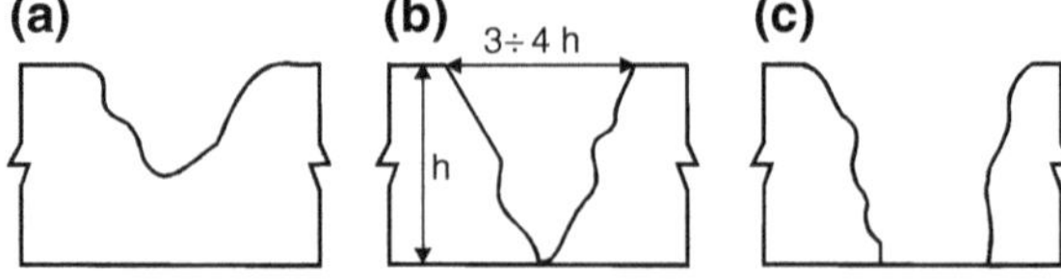

Fig. 2 Changes of breach shapes in time, created after overtopping (Johnson and Illes 1978): **a** initial breach shape; **b** "V"-shape breach, where h—breach depth; **c** final breach shape

The length of the front dam is 810 m while the width of the crest is 9 m, and the slope equal to 1:2.75. The spillway of the reservoir is a three-span weir of the following width: 3×6.0 m and equipped with three orifices of 1 m diameter. The parameters of the dam were designed according to the regulations for the first class of hydraulic structures, that considered the design of the capacity of the reservoir for 200 years period return flow ($Q_{0.5\%} = 115$ m^3/s) and 1,000 year period return flow ($Q_{0.1\%} = 136$ m^3/s). The control gates of the weir are 3 deflector gates, which lift water to the height of 2.70 m.

2.2 The Scenario of the Siemanowka Reservoir Earth Dam Failure

Relatively few failures of earth dams are considered to be very dangerous, because they initiate the erosion of the body of the dam and they lead to increasing of the breach in the dam. The way in which the dam becomes destroyed determines the characteristics of water outflow from reservoir. The most frequent cause for the breakdown of the dam is overtopping dam failure (Londe 1981; Wang and Bowles 2005). If an overtopping is observed, the water level in the reservoir must exceed the top of the dam before any erosion occurs. The size of the dam breach, which is then created, depends on the duration and the discharge rate of water flow over the top of dam.

If the overtopping failure is not significant and its duration is relatively short, then the breach is of the shape as shown in Fig. 2a. In the case of high discharge and long duration, the shape of the breach is of the forms shown in Fig. 2b and c.

Usually the highest discharge rate after overtopping failure occurs in the middle of the dam, and exactly in this place depth of the breach is the highest. Initially, the breach takes a V-shape, of the width 3–4 times higher than the depth. While the erosion of the dam progresses, the bottom of the breach becomes round, and the length of the breach depends on the size of the reservoir and the stored volume of water. This means that for large reservoirs there is the possibility of a complete erosion of the dam.

For the Siemianówka Reservoir it was assumed that damage to the gearing of the weir control gates occurs, which makes it impossible to provide capacity for

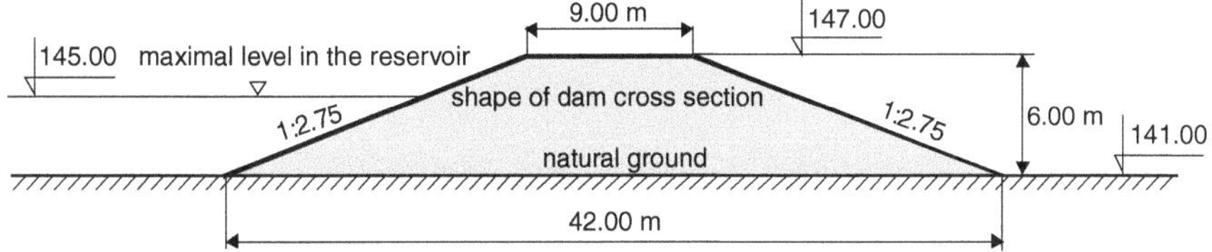

Fig. 3 Cross section of the Siemianówka Reservoir dam

200 years period return flow ($Q_{0.5\%} = 115$ m^3/s). In consequence, the reservoir reaches its full capacity and water flows over the crest of the dam. In the same time, the sluice and the spillway provide the total discharge equal to 77 m^3/s. The processes of the crest erosion and the breach forming are initiated. After overtopping failure, the discharge of water flowing into the Siemianówka Reservoir decreases to $Q = 77$ m^3/s during 70 h, which is twice the value of the long-term mean of the highest observed discharges. The reservoir surface area of the Siemianówka Reservoir is 3,250 ha if the water level reaches the top of dam altitude, equal to 147.00 m a.s.l.. The cross section of the Siemianówka Reservoir earth dam is presented on Fig. 3.

2.3 The Construction of the Breach Discharge Hydrograph

Singh and Quiroga (1988) provided an interesting dimensionless solution to the process of breach forming. It was based on the assumption that the initial shape of the breach (triangular or rectangular) is known, and the discharge of water to the reservoir is constant. For the reservoir flow the continuity equation is valid. Mean velocity of water in the breach is described by the equation of water outflow through free weir. The changes of the breach size caused by erosion are proportional to the mean velocity of water in the breach, raised to power "n", and to the erosion coefficient constant for certain soil type in the body of the dam. This assumption is correct, because the erosion of the breach is directly proportional to shear stress and in consequence to the mean velocity of water in the breach. This assumption enables one to derive equations for the determination of the breach edge location during the emptying of the reservoir and the total time of the breach formation. The results obtained by means of those equations are similar to the results of physical modeling of a dam washout process and to the results of mathematical modeling, which are nowadays considered to be the most reliable for such forecasts (Macchione 2008; Singh and Scarlatos 1988). Since the derived equations provide the results which are comparable with other methods, they were used in this paper to forecast the breach forming. All sufficient derivation of applied equations are based on parameters given by Kubrak and Szydłowski (2002, 2003), Szydłowski (2003) or genuine papers by Singh and Quiroga (1988).

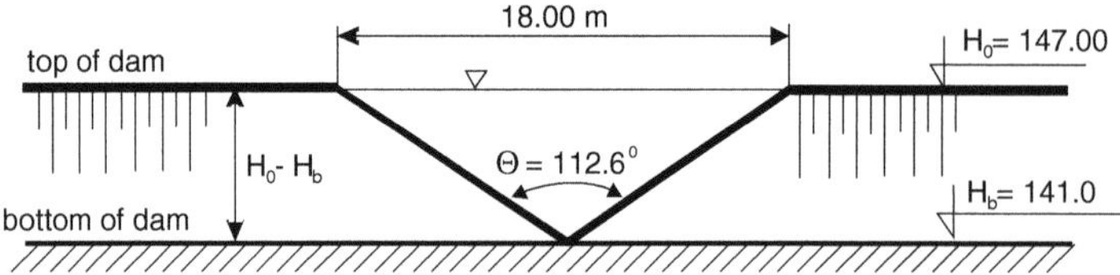

Fig. 4 The scheme for the calculation of triangular breach forming time by method of Singh and Quiroga (1988)

The hydrograph of water outflow from the Siemianówka Reservoir was constructed assuming that as a result of overtopping a triangular breach of a length 18.0 m will be created in the dam, and the lower vertex of the breach will be lowering until it reaches the bottom of dam. The time in which the breach will reach the bottom of dam is called the time of breach formation. The initial width of the breach was determined by applying statistical relationships, derived from the analysis of previous earth dam catastrophes (Froehlich 2008; Von Thun and Gillette 1990; Wahl 1998, 2004). Further development of the breach was not analyzed, because for the forecast of hydraulic consequences of celerity in the valley the highest values of water stages and discharges occur during the initial phase of breach formation.

For the calculation of the time of breach formation in the earth dam of the Siemanowka Reservoir by the method of Singh and Quiroga (1988), the following initial data was assumed:

- a triangular breach of the length equal to 18.00 m is created in the earth dam as a result of overtopping, and the lower vertex top of the breach,
- after the washout of the dam, reaches the bottom of dam $H_b = 141.00$ m a.s.l.,
- top of dam $H_o = 147.00$ m a.s.l.,
- reservoir surfaces area at water surface elevation from a reference datum $H_o = 147.00$ m a.s.l. $-A_s = 32.5$ km^2,
- dam soil erodiblity factor $E_f = 0.00042$ (according to Singh and Quiroga 1988),
- exponent in the dimensionless solution by Singh $n = 2$.

The conditions for solving the equations of the breach forming time were schematically presented in Fig. 4.

Based on the assumed data and an adopted length of the triangular breach base $B = 18$ m, the time of the breach formation was calculated, equal to $T_f = 6,360$ s = 106 min. The relation of the triangular breach top changes in the time was presented in Fig. 5.

As it can be concluded from Fig. 4, the top of the triangular breach, initially set at the altitude of 147.00 m a.s.l., reaches the bottom of dam at 141.00 m a.s.l. in 106 min.

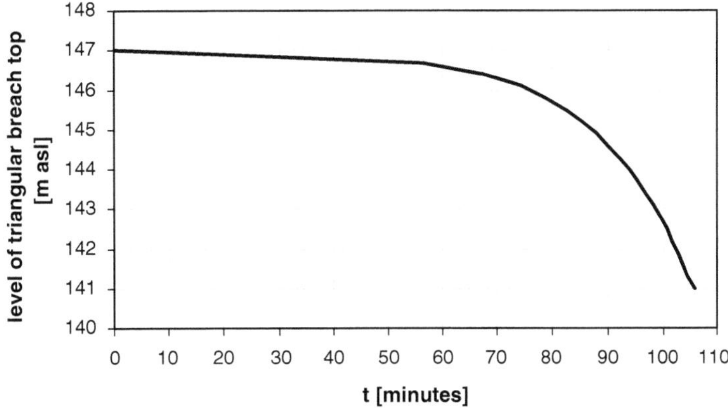

Fig. 5 The forecast of the lower vertex of the breach top elevation in time, calculated by method of Singh and Quiroga (1988)

2.4 Mathematical Description of Water Flow Created by the Outflow Through the Breach

The unsteady water flow in the reservoir and in the valley downstream is described by a set of one-dimensional, differential de Saint–Venant equations: continuity equation (conservation of mass):

$$\frac{\partial A}{\partial t} + \frac{\partial Q}{\partial x} = 0 \tag{1}$$

dynamics equation (conservation of momentum):

$$\frac{\partial Q}{\partial t} + \frac{\partial}{\partial x}\left(\frac{Q^2}{A}\right) + gA\frac{\partial h}{\partial x} + gAS_f = 0 \tag{2}$$

where H, water level elevation in a cross-section; A, cross-section area for water level H; g, gravity acceleration; S_f, friction loss; hydraulic gradient, Q, discharge; x, distance; t, time.

The equations of unsteady water flow (1), (2) were solved by applying a 4-point implicit scheme with a weighting parameter by Preismann (Cunge et al. 1980). This scheme approves to set the time step value as variable in a wide range, which is essential for the calculations performed for natural riverbeds and reservoirs.

In order to solve the nonlinear set of Eqs. (1) and (2) a general Newtonian method was applied, which, as an iterative method, requires the researcher to determine an initial approximation of the solution.

The initial conditions for the description of a wave motion are known values of water levels and discharges at all cross-sections of the Siemianówka Reservoir and the Narew River section, downstream of the reservoir at time $t = 0$. For the description of the ascending wave in the reservoir, the upstream boundary

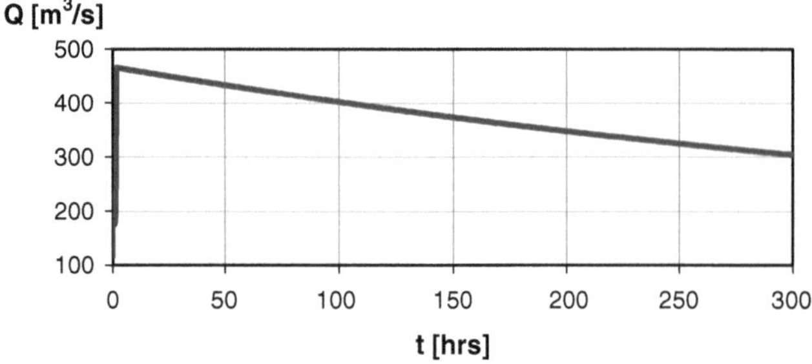

Fig. 6 Predicted discharge hydrographs for the triangular breach in the earth dam of the Siemianówka Reservoir overtopping failure

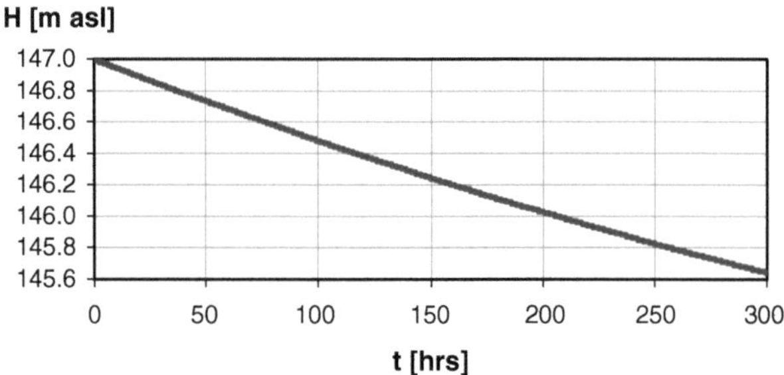

Fig. 7 Predicted stage hydrographs for the triangular breach in the earth dam of the Siemianówka Reservoir overtopping failure

condition (for the first, most upstream cross-section of the Siemianówka Reservoir) was defined as the discharge of water inflowing to the reservoir. The downstream boundary condition of the reservoir, located in the cross-section containing the breach, was the critical motion equation. The description of celerity in the valley downstream of the dam involved the upstream boundary condition (for the first cross-section in the Narew Valley) which was the previously calculated discharge of the breach and the discharge of both: the spillway and the sluice. The downstream boundary condition, for the most downstream cross-section of the analyzed reach of the Narew River, was that cross-section rating curve $Q = f(H)$.

3 The Calculation Results and Discussion

3.1 The Results of the Dam Break Wave Parameters Calculation

Using previously calculated temporal changes of the elevation of the triangular breach top, the one dimensional equation of unsteady water flow (1), (2) in the Siemianówka Reservoir was solved, applying the critical flow equation in the breach

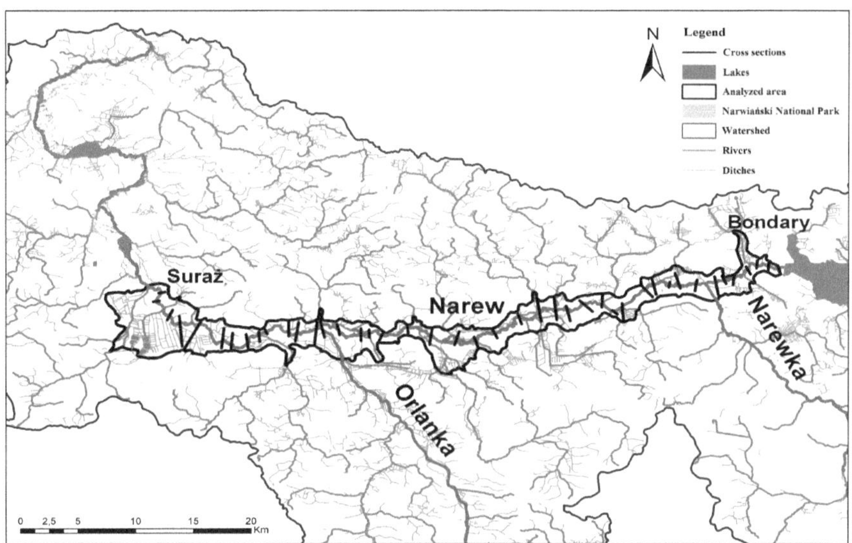

Fig. 8 Location of cross-sections used for calculations

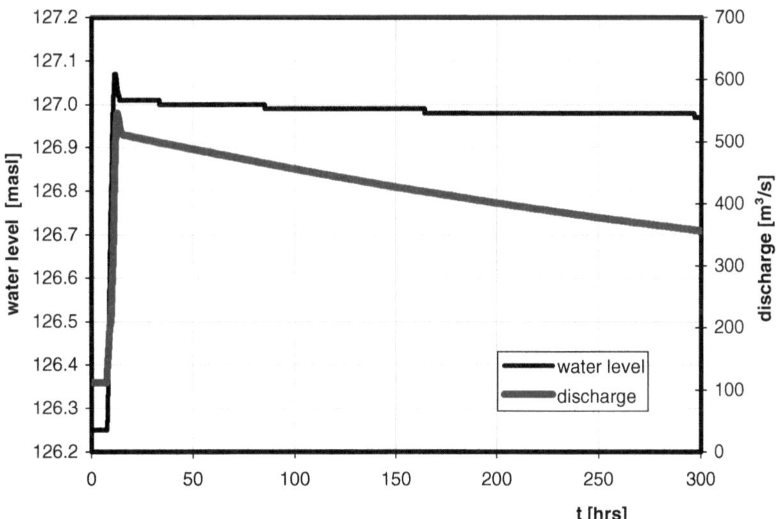

Fig. 9 Exemplary predicted stage and discharge hydrograph for the Narew Valley cross-section at 37.450 km, during the flood wave occurrence after the dam break

as a downstream boundary condition. Moreover, it was assumed that the breach cross-section is not submerged, but free. The upstream boundary condition in the reservoir was the inflow discharge value, derived from the falling limb of the flood discharge hydrograph. The hydrograph was constructed, assuming that a constant

Table 1 The initial (H_o), maximum calculated stages (H_{max}), discharges (Q_{max}) and the time of occurrence in the Narew River after the failure of the Siemianówka Reservoir earth dam

Cross-section	Km downstream of the dam	H_o (m a.s.l.)	H_{max} (m a.s.l.)	T_{Hmax} (h)	$H_{max} - H_o$ (m)	Q_{max} (m³/s)	T_{Qmax} (h)
P60	0.000	137.92	138.38	1.17	0.46	466	1.77
P59	2.250	137.52	137.99	5.33	0.47	478	6.27
P58	3.875	137.44	137.85	5.70	0.41	481	6.07
P57	5.875	136.60	136.97	6.10	0.37	489	6.13
P55	10.375	134.32	134.70	6.67	0.38	533	6.27
P51	13.375	133.24	133.76	7.90	0.52	533	6.33
P49	17.500	133.14	133.24	8.50	0.10	547	6.40
P44	18.575	131.91	133.04	8.87	1.13	647	6.43
P43	19.825	130.94	131.73	9.73	0.79	708	6.47
P42	20.825	130.25	131.56	9.87	1.31	575	6.63
P41	21.825	129.98	130.69	10.80	0.71	526	6.90
P40	23.200	129.75	130.46	11.30	0.71	513	11.53
P39	24.675	129.69	130.40	11.33	0.71	513	11.60
P38	25.925	129.15	129.67	12.00	0.52	513	11.73
P37	27.425	128.44	129.08	12.40	0.64	513	12.10
P36	29.175	128.09	128.71	12.53	0.62	513	12.47
P35	30.675	127.53	128.30	12.73	0.77	512	12.87
P34	32.425	126.99	127.70	12.87	0.71	513	12.90
P33	35.800	126.40	127.14	12.90	0.74	530	12.97
P31	37.450	126.17	127.07	12.93	0.90	546	13.03
P30	38.875	125.76	126.41	13.20	0.65	549	13.17
P29	40.550	125.14	125.80	13.57	0.66	544	13.33
P28	42.425	124.76	125.41	14.03	0.65	538	13.67
P27	44.300	124.56	125.22	14.17	0.66	538	14.13
P23	45.500	124.52	125.04	14.20	0.52	535	14.23
P21	47.500	123.73	124.21	14.50	0.48	544	14.50
P3	49.750	122.79	123.38	15.17	0.59	578	15.03
P17	51.500	122.69	123.32	15.37	0.63	573	16.43
P15	54.375	122.66	123.27	16.30	0.61	572	16.57
P2	55.375	122.27	123.26	16.43	0.99	561	16.73
P13	57.625	121.29	123.25	17.83	1.96	550	17.47
P12	59.250	121.12	123.25	17.97	2.13	545	17.57
P1	60.750	120.70	123.24	18.00	2.54	532	17.60
P11	61.750	120.30	123.24	18.03	2.94	521	17.77
P8	66.000	119.36	123.24	18.13	3.88	506	18.67
P6	68.625	119.25	123.24	19.77	3.99	458	19.27

inflow of water to the reservoir is equal to 200 years period return flow ($Q_{0.5\%} = 115$ m³/s) until the water spill over the crest. Then, the discharge decreases within 70 h to the value of $Q = 77$ m³/s, which is twice the value of a long-term mean of the highest observed discharges. The Siemianówka Reservoir was described by 9 cross-sections, and the time step of the calculations was equal to $\Delta t = 120$ s.

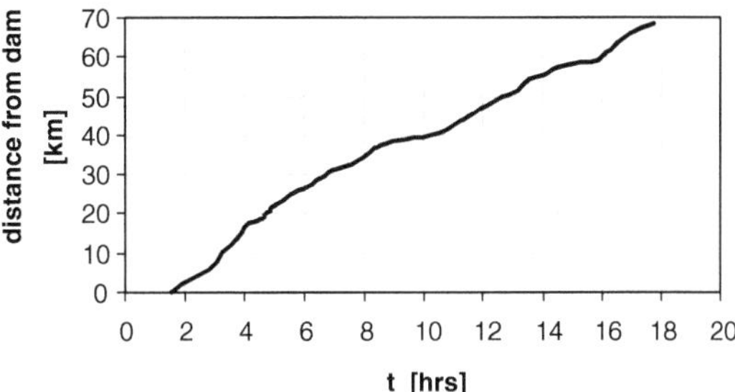

Fig. 10 Predicted wave-front propagation in the Narew River valley after the overtopping failure of the Siemianówka Reservoir dam

Figures 6 and 7 presents the calculated discharge and stage hydrographs for the triangular breach in the dam of the Siemianówka Reservoir.

3.2 The Results of Flood Wave Parameters Calculation for the Valley of the Narew River after the Siemianówka Dam Break

For the calculation of flood wave propagation in the Narew Valley, 36 cross-sections were used, which describe a 68-km section of the riverbed (Fig. 8). Discharge hydrograph of the breach, shown in Fig. 6, was used as the upstream boundary condition for the calculations of flood wave propagation in the valley of the Narew River. The analyses were performed for a 68.625 km long section of the riverbed and the valley, reaching from the Siemianówka Reservoir dam to the Suraż cross-section and described by 36 cross-sections (Fig. 8). Constant inflows of water to the Narew River from the Narewka River and the Orlanka River were assumed, respectively: 49.2 and 53.6 m^3/s. On the base of the flood wave calculations for the Narew Valley, discharge and stage hydrographs were constructed for each cross-section (Fig. 9). The maximum calculated water stages and discharges for individual cross-sections and their duration have the greatest significance for practical purposes. The maximum calculated stages and discharges as well as the time of occurrence are presented in Table 1. That table also contains water stages in the Narew River (H_o) before the occurrence of flood wave, caused by the dam break.

The determination of wave-front occurrence time at individual cross-sections of the valley is also essential for the planning of flood protection measures. The calculation results were used to determine the propagation of wave-front in the valley. The wave-front occurrence was considered to be the moment of the raise of water levels by 0.2 m. higher than before the dam break. The results were presented in Fig. 10.

The inundation areas in the Narew Valley, after the overtopping failure of the dam of the Siemianówka Reservoir, are shown on Figs. 11, 12, 13, and 14.

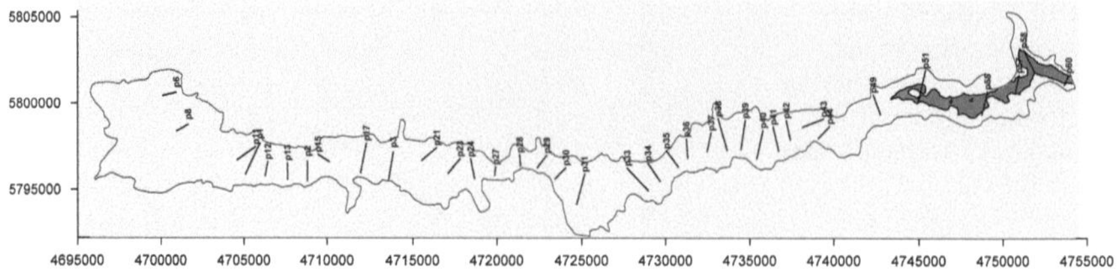

Fig. 11 Inundation area 4 h after the Siemianówka Reservoir dam overtopping failure

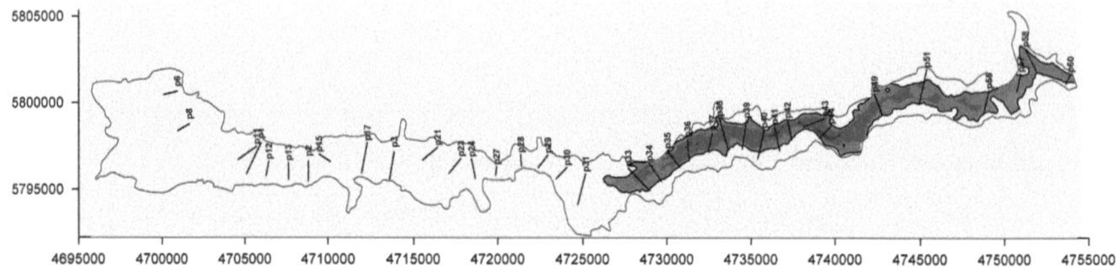

Fig. 12 Inundation area 8 h after the Siemianówka Reservoir dam overtopping failure

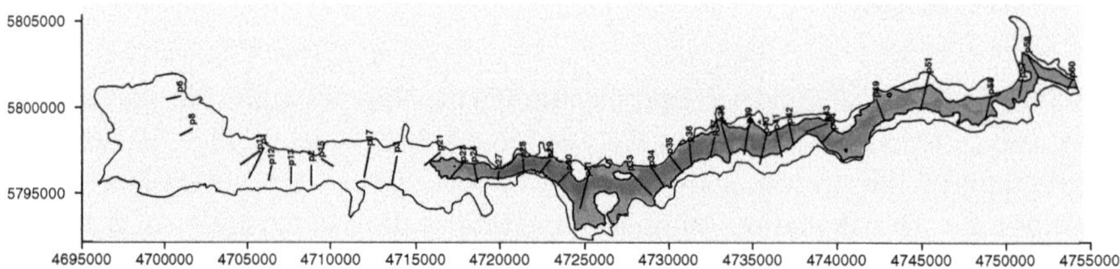

Fig. 13 Inundation area 12 h after the Siemianówka Reservoir dam overtopping failure

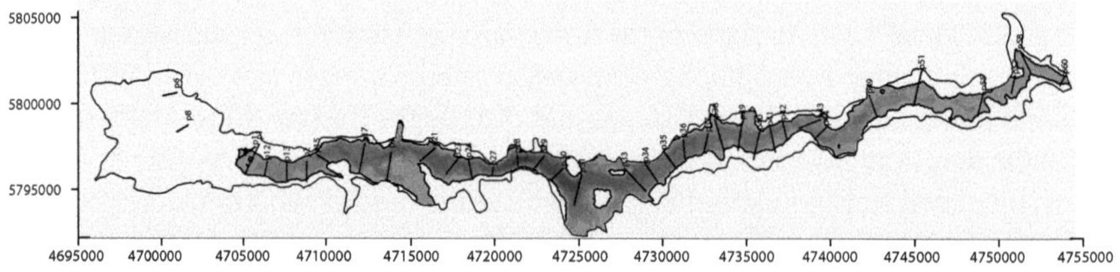

Fig. 14 Inundation area 16 h after the Siemianówka Reservoir dam overtopping failure

4 Conclusions

A low probability of the dam break does not mean, however, that it is completely
unreal. The threat of the dam break, which suddenly releases the mass of stored
water, leads to such a discharge rate of water from the reservoir that can be hardly

compared to the threats presented by natural floods. Numerical analyses of the flood wave propagation in the valley downstream of the dam enable us to elaborate evacuation and emergency plans, warning people of the potential flood threat. Information provided by numerical simulations is extremely vital for the administration of the areas threatened by flooding. The presented example of calculations for the Siemianówka Reservoir underlines the most important issues related with the forecast of the dam break, that is: the determination of maximum water levels and discharges in the valley downstream of the failure dam, the calculation of wave-front occurrence time at particular points of the valley, the inundation range along with the longitudinal profiles of water levels and velocities of water on inundated areas. The knowledge of the water depths on inundated areas, the duration of inundation and maximum water velocities enable as well the estimation of the damage degree on inundated areas. As it results from the calculations, the washout process of the Siemianówka Reservoir earth dam, which is 6 m high, happens very quickly, and after about 2 h the top of the triangular breach reaches the dam bed. The discharge from the reservoir after dam failure drops from 465 to 304 m^3/s and the water level in reservoir decreases 1.36 m as short as within 300 h. The Narew valley downstream of the Siemianówka Reservoir is broad and covered by peatlands. The occurrence of wave in the valley causes the increase of water levels by 0.1–4 m in comparison to the water level before the dam break, and a 4-time increase of discharge rate in relation to the design flow at the cross-section of the dam of the Siemianówka Reservoir. The velocity of the wave-front propagation in the Narew Valley does not exceed 5 km/h.

References

Arico C, Nasello C, Tucciarelli T (2007) A Marching in Space and Time (MAST) solver of the shallow water equations, Part II: the 2-D model. Adv Water Res 30(5):1253–1271

Chanson H (2009) Application of the method of characteristics to the dam break wave problem. J Hydraul Res 47(1):41–49

Cunge JA, Holly FM Jr, Verwey A (1980) Practical aspects of computational river hydraulics. Pitman, London

De Wrachien D, Mambretti S (2009) Dam-break problems, solutions and case studies. WIT Press, Southampton

Fread DL (1977) The development and testing of a dam-break flood forecasting model. Proceedings of dam-break flood routing model workshop, US Department of Commerce, National Technical Information Service

Fread DL (1984) DAMBRK. The NWS dam-break flood forecasting model. National Weather Service, Maryland

Fread DL (1985) BREACH, an erosion model for earthen dam failures. NWS Report, Oceanic and Atmospheric Administration, Silver Spring

Froehlich D (2008) Embankment dam breach parameters and their uncertainties. J Hydraul Eng 134(12):1708–1721

Gallegos H, Schubert J, Sanders B (2009) Two-dimensional, high-resolution modeling of urban dam-break flooding: a case study of Baldwin Hills, California. Adv Water Resour, pp 1324–1335

Johnson FA, Illes P (1978) A classification of dam failures. Water Power Dam Constr 28:43–45

Kubrak J (1989) Numerical models for the analysis of dame breach flood wave. Treatises and Monographs. WULS, Warsaw (in Polish)

Kubrak J, Szydłowski M (2002) Application of Singh and Quiroga model for the forecast of the discharge hydrograph through a rectangular breach in the earth dam. Sci Rev Eng Environ Sci, WULS, vol 1, no 24, pp 240–250 (in Polish)

Kubrak J, Szydłowski M (2003) Application of parametrical models for the determination of the earth dam breach parameters. Sci Rev Eng Environ Sci, WULS, vol 1, no 26, pp 52–60 (in Polish)

Londe P (1981) Conclusions from earth dam breakdowns. Gospodarka Wodna, nr 7–8

Macchione F (2008) Model for predicting floods due to earthen dam breaching. I: formulation and evaluation. J Hydraul Eng ASCE 134(12):1688–1696

Singh VP, Quiroga AC (1988) Dimensional analytical solutions for dam-break erosion. J Hydraul Res 26(2):179–197

Singh VP, Scarlatos PD (1988) Analysis of gradual earth-dam failure. J Hydraul Eng ASCE 114(1):21–42

Soarez Frazão S, Zech Y (2002) Dam Break in channels with 90° bend. J Hydraul Eng ASCE 128(11):956–968

Sokolowski J (1999) The Siemianówka Reservoir monograph. Voivodship Board of Land Reclamation and Hydraulic Structures, Bialystok (in Polish)

Szydłowski M (1998) Numerical simulation of water flow with a free surface in the conditions of rapidly-changing, unsteady motion with discontinuities. Ph.D. Dissertation, Gdansk University of Technology

Szydłowski M (ed) (2003) Mathematical modeling hydraulic impact failure of water dams. Monographs Committee on Water Resources Management of Polish Academy of Sciences, Warsaw (in Polish)

Von Thun JL, Gillette DR (1990) Guidance on breach parameters. U.S. Bureau of Reclamation, Denver

Wahl TL (1998) Prediction of embankment dam breach parameters: literature review and needs assessment. U.S. Bureau of Reclamation Dam Safety Report DSO-98-004

Wahl TL (2004) Uncertainty of predictions of embankment dam breach parameters. J Hydraul Eng 130:389–397

Wang Z, Bowles DS (2005) Three-dimensional non-cohesive earthen dam breach model. Part 1, theory and methodology. Adv Water Resour 11.009

Index

D. Mirosław-Świątek and T. Okruszko (eds.), *Modelling of Hydrological Processes in the Narew Catchment*, Geoplanet: Earth and Planetary Sciences, DOI: 10.1007/978-3-642-19059-9, © Springer-Verlag Berlin Heidelberg 2011

MIX
Papier aus verantwortungsvollen Quellen
Paper from responsible sources
FSC® C105338

If you have any concerns about our products,
you can contact us on
ProductSafety@springernature.com

In case Publisher is established outside the EU,
the EU authorized representative is:
Springer Nature Customer Service Center GmbH
Europaplatz 3, 69115 Heidelberg, Germany

Printed by Libri Plureos GmbH
in Hamburg, Germany